Exploring Vancouver Naturehoods

Vicky Earle

Exploring Vancouver Naturehoods

An Artist's Sketchbook Journal

Midtown Press

Midtown Press
7375 Granville Street
Vancouver, BC V6P 4Y3
Canada
www.midtownpress.ca

ISBN 978-1-988242-49-1 (Hardcover)
ISBN 978-1-988242-48-4 (Softcover)

Legal deposit: 1st quarter 2023
Library and Archives Canada

Printed and bound in China

Editor: Louis Anctil
Page layout: Vicky Earle
Typesetting and production: Denis Hunter Design

Library and Archives Canada Cataloguing in Publication

Title: Exploring Vancouver naturehoods : an artist's sketchbook journal / Vicky Earle.
Names: Earle, Vicky (Artist), author, artist.
Identifiers: Canadiana 20220483272 | ISBN 9781988242491 (hardcover) | ISBN 9781988242484 (softcover)
Subjects: LCSH: Earle, Vicky (Artist)—Notebooks, sketchbooks, etc. | LCSH: Nature in art. | LCSH: Biodiversity—British Columbia—Vancouver. | LCSH: Vancouver (B.C.)—In art.
Classification: LCC NC143.E27 A4 2023 | DDC 741.971—dc23

Contents

Nature All Around Us

The extraordinary biodiversity in our city comes into view when seen through the pages of an artist's sketchbook journal. *Exploring Vancouver Naturehoods* takes a closer look at nature wherever you find it in the city – in beautiful parks, along tree-lined streets, amid hedges by a building, or peeking through cracks in a sidewalk. These are Vancouver's "naturehoods" – any green space in your neighbourhood, no matter how big or how small. These are places to connect with nature's stories on a personal level, whether listening to a bird's song, watching a dragonfly hunt for lunch, or musing at the antics of a squirrel. Using a sketchbook as a tool helps us slow down and interact more meaningfully with nature. It can also be a portal that connects us with community on many levels.

It's fascinating how an artist with a sketchbook, looking intently at something, opens conversations with interested passersby. I love how sketching outdoors piques people's curiosity. My sketchbook journal has become a great way to connect with others in the community, to share the stories of what I'm looking at, and to help neighbours take a closer look at nature. Seeing aha moments take shape right before my eyes makes that sketching session even more worthwhile.

Our communities are made up of more than just "us." City residents include the butterflies that arrive in spring, the flowers along boulevards, and the bald eagles that soar overhead. Our city also hosts a vast number of nature "tourists" – species that come here either as a seasonal destination to nest and raise their young or as a resting place during a much longer journey. British Columbia is part of the Pacific Flyway, a major migratory route stretching from Alaska to Patagonia. Spring and fall in Vancouver are the times to see snow geese, rufous hummingbirds, all types of swallows and warblers, and a variety of shorebirds and ducks as they pass through the city. The change of seasons brings fascinating transitions in plant and insect life, showcasing a myriad of colours, shapes, patterns, and textures. Infinitely complex and interconnected, nature provides something remarkable waiting to be discovered with every outing. The nature sketches in *Exploring Vancouver Naturehoods* only scratch the surface of the natural landscape and wildlife that can be discovered in our local green spaces.

There is increasing concern that urban residents around the world – children and youth (especially) – are spending less time outdoors and losing our connection with nature and the health benefits it provides. This book encourages city residents of any age to enjoy rich experiences with nature everywhere we find it. That could be a spider in its web at the bus stop or an osprey bringing fish to its chicks on the Fraser River. Everything has a story. Watching nature brings a sense of calm and restores balance to our spirit. A growing body of research shows that spending time in nature – even as little as twenty minutes – benefits our mental and physical health. Some doctors, in partnership with the Parks Prescription Program, or PaRx, have begun to prescribe time in green space for their patients. Positive health effects are many – from reducing chronic disease (like asthma, diabetes, heart disease, high blood pressure, and stroke) and lowering blood pressure and stress hormones in adults, to improving eyesight and brain power in kids – all over and above the benefits of exercise! Even more impressive, adults who take short day trips to the forest boost their levels of immunoproteins and natural killer cells (a type of white blood cell), improving immune functions for at least seven days after the trip. Memory, creativity, and task performance also increase after a walk in a park.

Keeping a nature sketchbook – or nature journaling – is a way to fall in love with nature. A nature journal is a simple, accessible tool available to everyone. Keeping a journal is intimate and engaging – it is a personal exploration of the nature around us. Experiences become more memorable than snapping a quick pic on a smart phone and moving on (although that has its place too). Nature journaling is a way to appreciate nature more deeply, to be truly mindful, and to have fun with creativity. The more we engage with nature through a sketchbook, the more nature seemingly engages with us. When we are still and intently observing, wildlife becomes more comfortable with our presence. Often, a variety of creatures come flooding into view – almost like things *want* to be noticed. On more than one occasion, a bird or butterfly has brushed my head and landed in front of me just begging to be drawn.

Asking questions and letting curiosity be our guide is at the heart of keeping a journal. John Muir Laws, artist, naturalist, author, and educator, encourages journalers to keep three guiding principles in mind – "I notice, I wonder, and it reminds me of" – followed by recording information on the pages of your sketchbook through drawing, writing notes, and asking questions. Many different parts of our brain are activated during this process. When done regularly, journaling impacts how we think and increases our ability to look with intention, make connections, and think visually. There is no better way to connect a person to a place and a place to a person.

The wonders of nature are all around us – we only need to look.

About This Book

This book grew from a desire to help people connect more meaningfully with nature. While using a sketchbook journal is just one way to do this, it can and hopefully will become a regular practice in people's lives.

Opportunities to connect with nature in Vancouver are endless. More than 250 parks and green spaces can be found within the city limits (plus twenty-two regional parks, five greenways, and three ecological reserves in Greater Vancouver). These parks offer a variety of natural habitats, including rivers, forests, meadows, tidal flats, marshlands, and the ocean – all supporting a unique diversity of life just waiting to be observed and sketched.

Each chapter focuses on wildlife and plants found in different locations, arranged from west to east across the city. The chapter Close to Home includes sketches of species encountered near the artist's home (but applicable anywhere in the city). Vancouver Parks highlights selected parks across Vancouver. Some of these green spaces are well known, for example, Stanley Park, Jericho Beach Park, and Pacific Spirit Park, while others might be considered hidden gems, like Hastings Park Sanctuary and Fraser River Park. Only green spaces that are free for the public to visit are included. The section Creating Your Own Nature Sketchbook is dedicated to helping readers find nature in their own neighbourhoods and shares tips on starting a nature sketchbook journal.

Original sketchbook journal pages are done in graphite, ink, and watercolour with notes describing Vancouver's more-than-human world: plants, animals, birds, and insects that thrive in our parks and green spaces. Together, these sketches are not an exhaustive account or a guidebook of species but a sharing of things that caught the artist's eye and brought joy to that outing. Colour tests are sometimes included on the page. These tests are valuable to work out what mix of colours are best to use for the subject and also as a reminder if the sketch cannot be completed in one sitting. Many pages of this book were done on the spot. Others – depending on weather and time – were started in the field and finished in the studio. Specific observations about subjects are noted in real-time on the pages, while more in-depth research is often completed at home and added to the journal (sometimes days or weeks) later. For example, pages on the Anna's hummingbird are a culmination of watching this bird's habits at the feeder and in the park over a few months.

This book is an invitation to go outside and connect with nature – to put "green dates" in your calendar and become an explorer in your own city too. You don't need to take a sketchbook, but it's more fun when you do! Grab a pencil and some paper, head outside, and see what you discover.

As the artist and author of this book, I am grateful to live, work, and create art on the traditional territory of the Coast Salish Nations, including xʷməθkʷəy̓əm (Musqueam), Stó:lō and Səl̓ílwətaʔ/Selilwitulh (Tsleil-Waututh), and Sḵwx̱wú7mesh (Squamish) who have stewarded this land since time immemorial. I acknowledge the traditional custodians of the land and pay respect to the Elders, past and present. While traditional stories of this land go back thousands of years, the park histories shared here are often of relatively recent developments. I leave it to others to reveal the complex and personal narrative of Vancouver's urban evolution.

Close to Home

Unexpected Visitor

Countless nature treasures, textures, and mini worlds can be found every day, right outside your front door. Venturing out after a storm or strong winds can increase the interesting things you find. The change of seasons also brings new and, occasionally, unusual mementos to discover on your outings. Even taking time to look closely at a "weed" can reveal incredible patterns, structures, and colours.

If you cannot get outside, taking a few moments to gaze out a window can reveal surprising things. The heron below (drawn twice) was a delightful sight on a dreary November day during the COVID-19 pandemic. As I looked out the kitchen window, this large, beautiful bird was a wonderful distraction during those months of isolation. As he sat preening on the chimney top next door, I quickly grabbed my sketchbook and pencils to capture his poses on the page.

Found Treasures

August 1 to 20, 2021

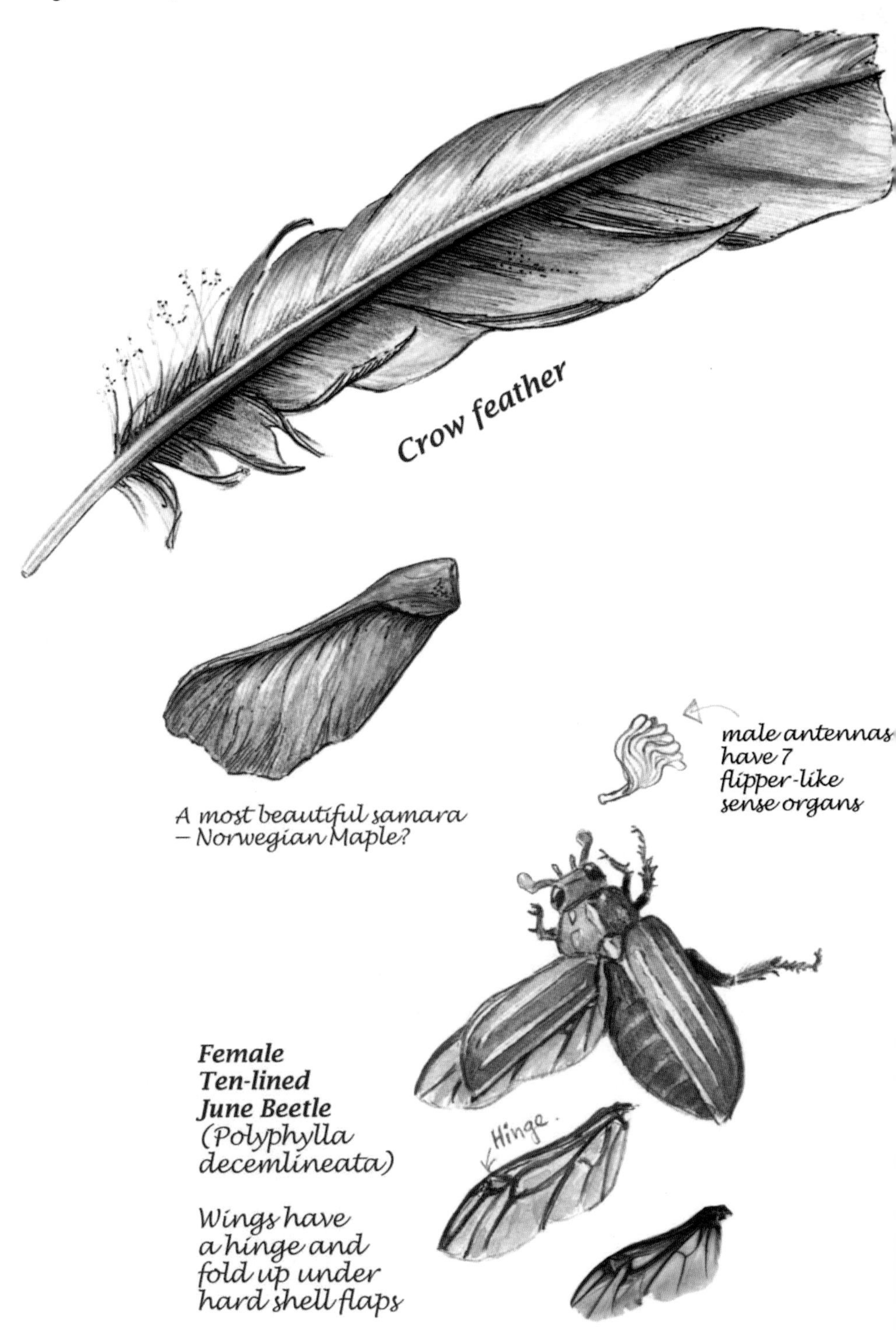

A walk through the neighbourhood after a heavy rain can reveal wonderful treasures – little bits of magic just beyond our doorstep – all with their own stories.

Sycamore tree seed balls made up of approximately one hundred individual achenes (hard, dry, single seeds) are a favourite food for beavers, muskrats, and squirrels.

I often find leaves, feathers of all types, and sometimes insects that have fulfilled their life duties, like the June beetle sketched here.

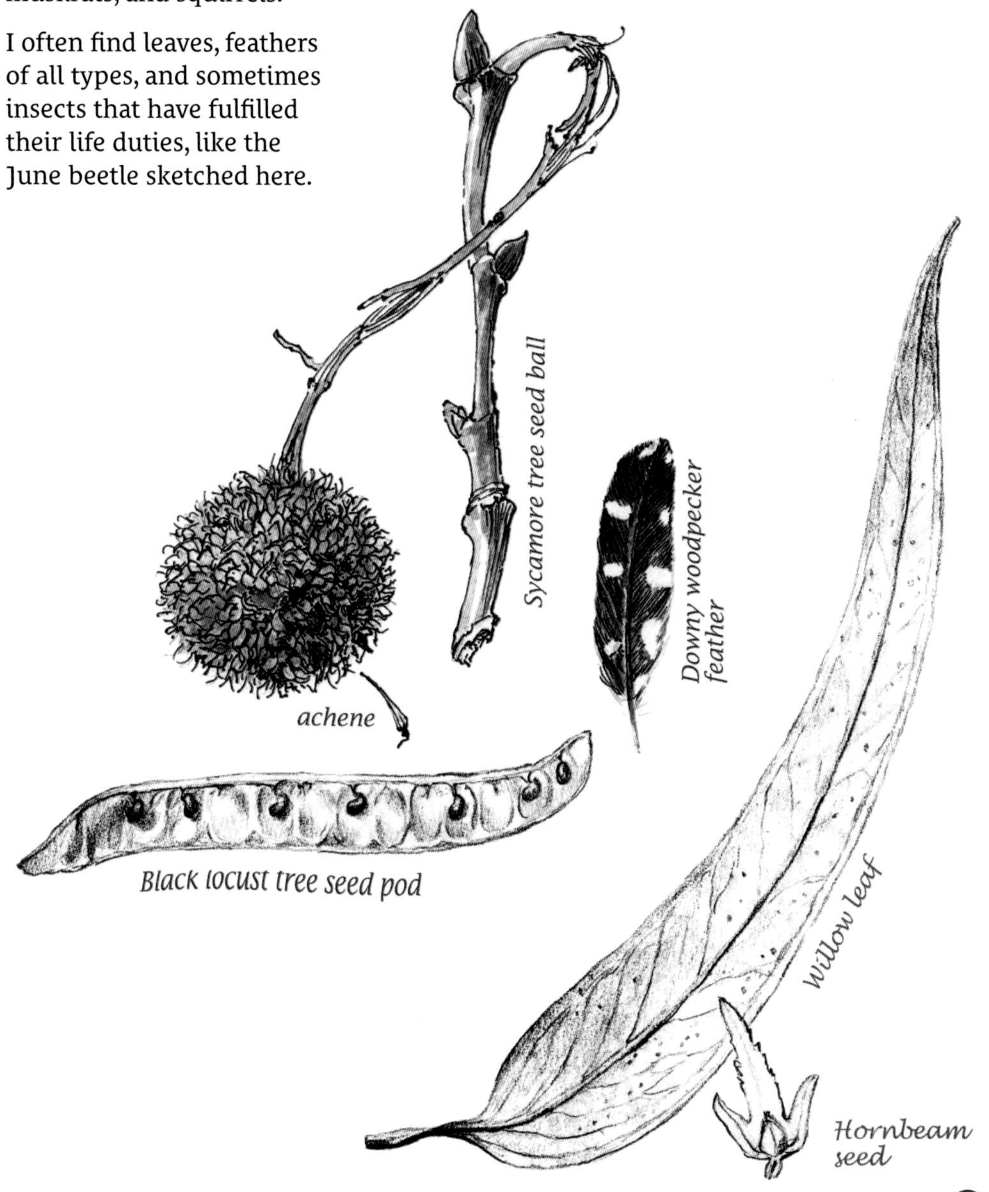

Limitless Leaf Shapes

April 5, 2020

Leaves come in a dazzling array of shapes, sizes, and textures. They are adapted for specific functions that the plant needs to survive in its environment.

Bumble Bee

(*Bombus suckleyi*)
on lavender bee balm
July 20, 2021

This bumble bee decided to take a nap in the bee balm on my patio, so I was able to get very close and take a good look at its face. To my delight, I discovered three additional eyes on the top of its head! Bees share this trait with dragonflies, grasshoppers, hornets, and wasps.

Exquisite Textures

various lichen species
seen throughout the year

Taking the time to get acquainted with the trees in your local neighbourhood often reveals tiny communities of lichens and mosses. Stop and take a closer look. What you find may surprise you! Lichens are complex life forms that are actually a partnership between a fungus and an alga. Lichens benefit us directly by their ability to absorb pollutants and are biological indicators of air quality in the environment. The colours, structures, and growing patterns of lichens also provide endless inspiration for an artist's sketchbook.

PIXIE CUP LICHEN

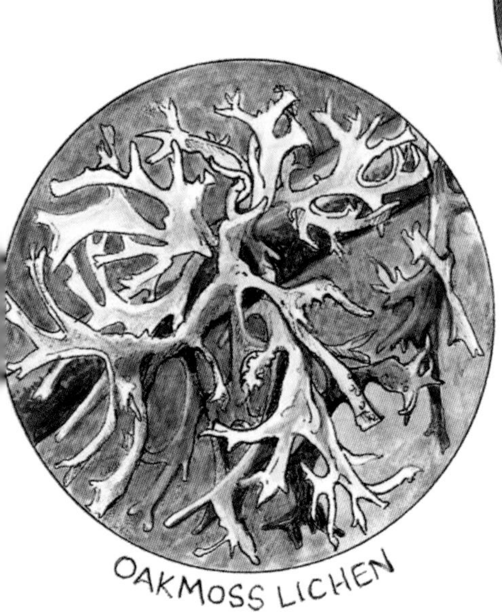

OAKMOSS LICHEN

GREEN-EYED ROCKBRIGHT

GREEN FOLIOSE & YELLOW CRUST LICHEN

A Day of Snails

(*Cepaea nemoralis*)
August 31, 2016

Considered by many to be pests, snails can be mesmerizing to watch and are important to a healthy ecosystem. On an August day after a heavy rain, I found at least eleven snails making their way along the sidewalk. I scooped up three and carried them to my studio where I put them in a clear container along with leafy twigs to draw.

Grove snails (*Cepaea nemoralis*) are quite colourful, ranging from yellow to light green to dark brown, with a lovely spiral band. This common snail was introduced here by European settlers and is fairly innocuous. Land snails eat material like fungus, rotting leaves, and algae – things that are close to the bottom of the food chain. In turn, snails become vital food for salamanders, snakes, birds, and mammals.

Studies have found that snails are such an important food source for some birds like song thrushes that, without them, these birds are unable to get enough calcium to lay eggs with strong shells and healthy embryos.

Brown snails (*Cornu aspersum*) are considered a garden menace because they outcompete other beneficial snails. Brown snails can reproduce once a month and survive up to minus 10 degrees Celsius.

Snails can see – they have eyes at the end of their tentacles. The lower tentacles on either side of their mouth are flexible and have smell and taste receptors on the tips. They stretch these tentacles forward as they move to help find food and safe nesting areas. Snails don't have ears, but they can sense vibrations that help them detect different sounds.

One of the largest terrestrial snails, the Oregon forest snail (*Allogona townsendiana*), is endangered. Its Canadian population is now restricted to southwest British Columbia and is rarely seen.

Day of Snails!
out on sidewalks on
a Rainy Day

Unusual Samaras

(*Acer pseudoplatanus*)
June 24, 2021

Maple seeds (samaras) are typically fused in twos that easily break apart after they helicopter to the ground. This sycamore maple (*Acer pseudoplatanus*) had many

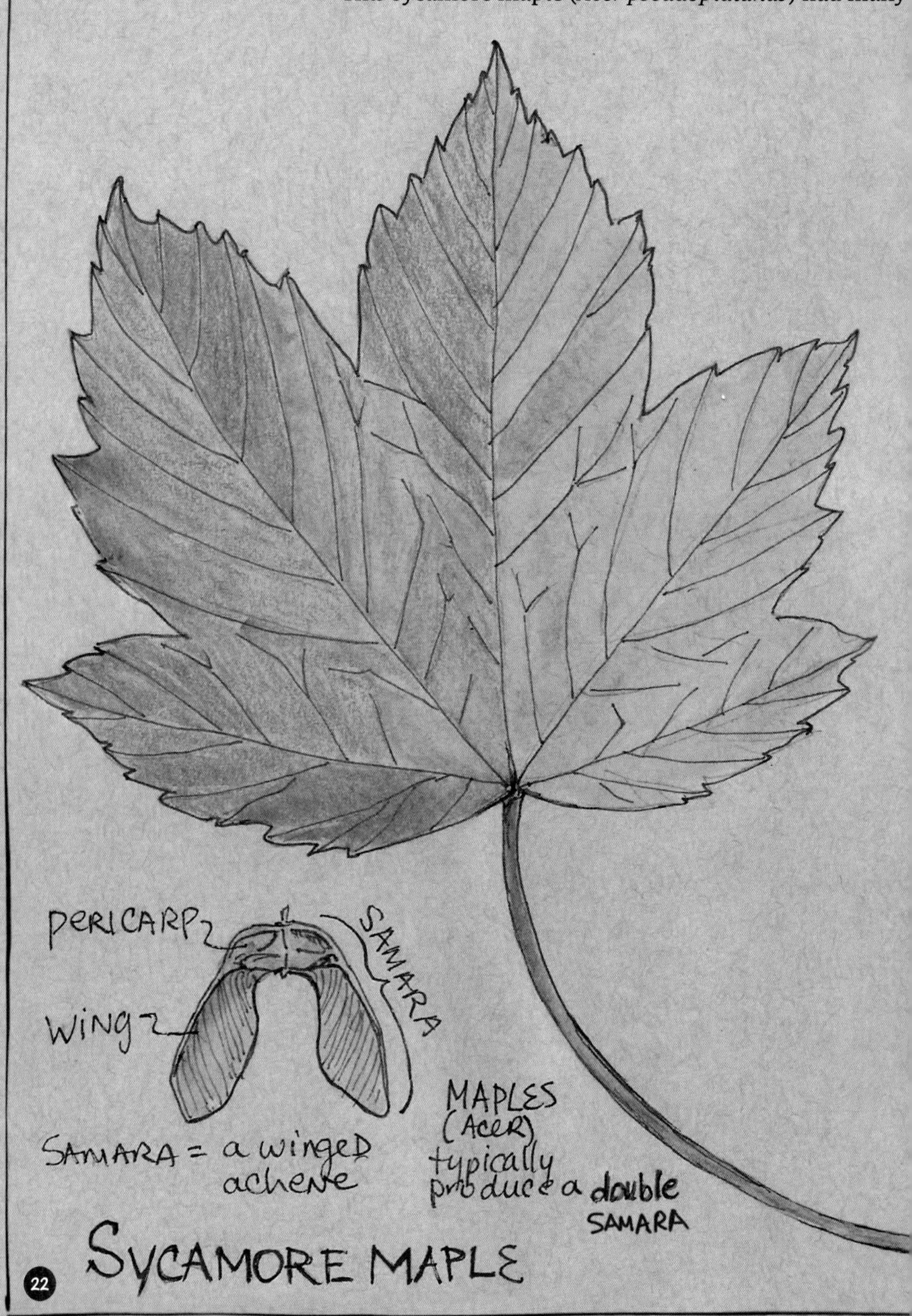

samaras fused into threes and fours – something I had never seen before. This curious find triggered many questions, listed below.

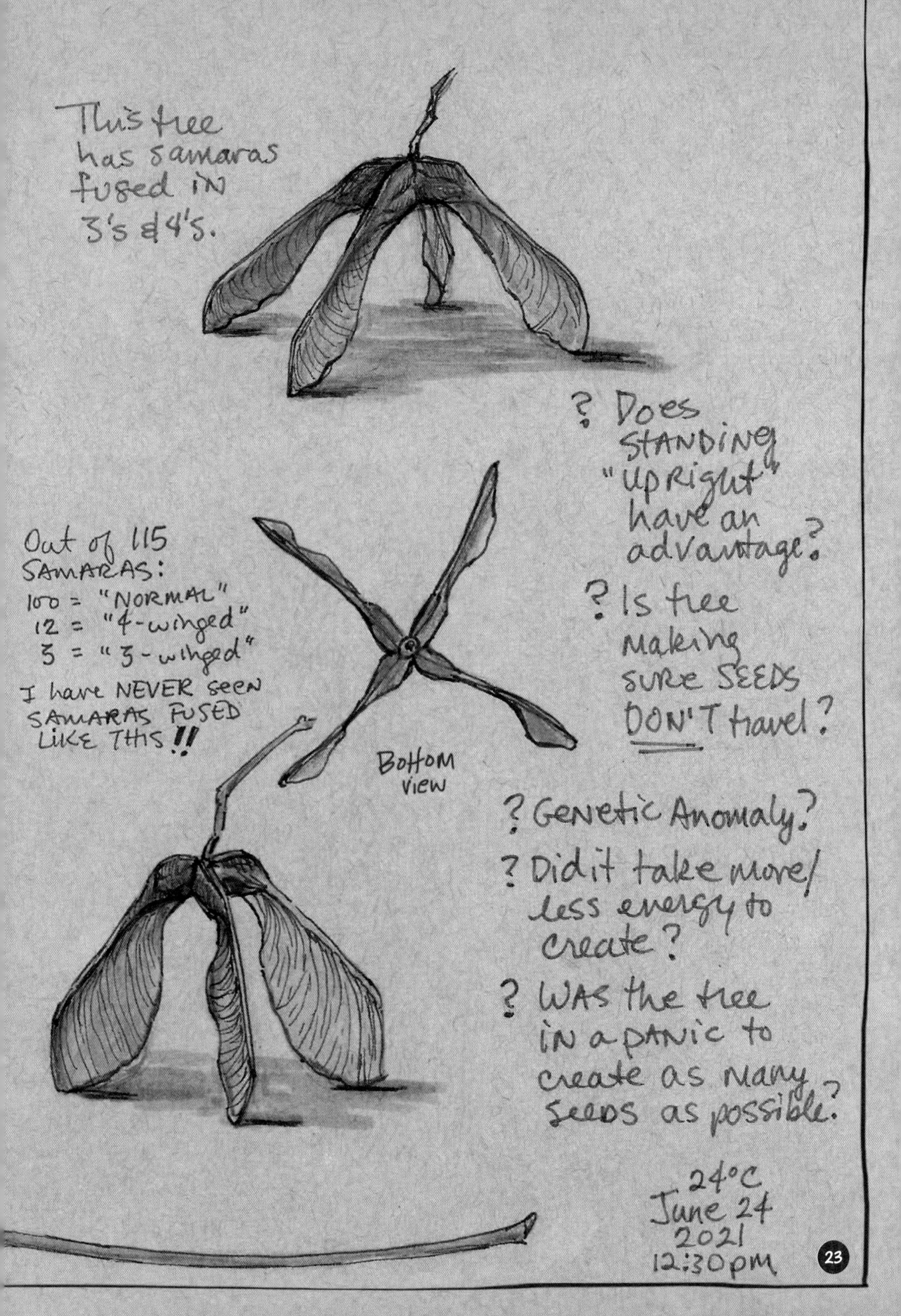

Walking through the Neighbourhood

March 13, 2021

When you open all of your senses, even a short walk around local streets reveals countless opportunities to get to know your nature neighbours and their personalities.

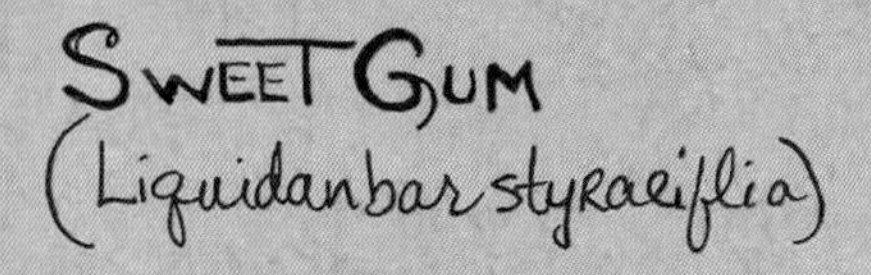

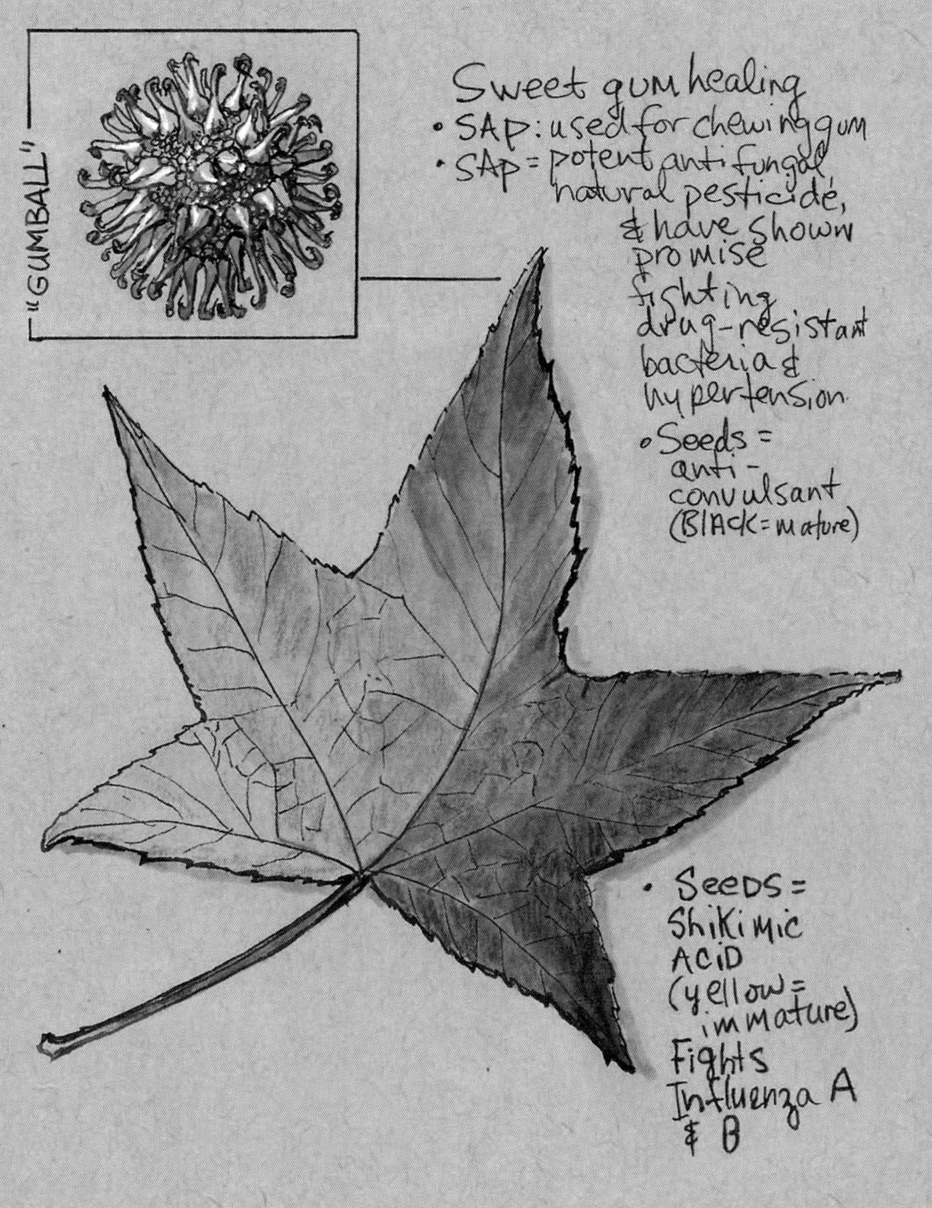

Colours, textures, patterns, scents, and sounds abound. A favourite, comforting sound is the "beep, beep, beep" call of the chestnut breasted nuthatch – a small, colourful delight on two legs that searches for insects, nuts, and seeds to dine on.

The sweet gum tree is an astonishing plant. Its seed pods, called "gum balls," can be found throughout the year. On close inspection, the "spikes" grow in pairs and remind me of a bird's beak. Once mature, the "beaks" open up, letting tiny seeds spill out. In the fall, the leaves change into a splendid burst of colours – sometimes showing green, yellow, oranges, reds, and deep burgundy at the same time, on the same tree.

Cooper's Hawk

(*Accipiter cooperii*)
March 1, 2021

This hawk took up residence outside my window one morning in early spring, much to the dismay of the local songbirds. Hawks typically bond for life and I later found their nest a few blocks away.

- FEMALES ARE LARGER THAN MALES
- FIELD I.D. = ROUNDED TAIL FEATHERS

8 AM
MARCH 1,
2021
OUTSIDE
MY WINDOW

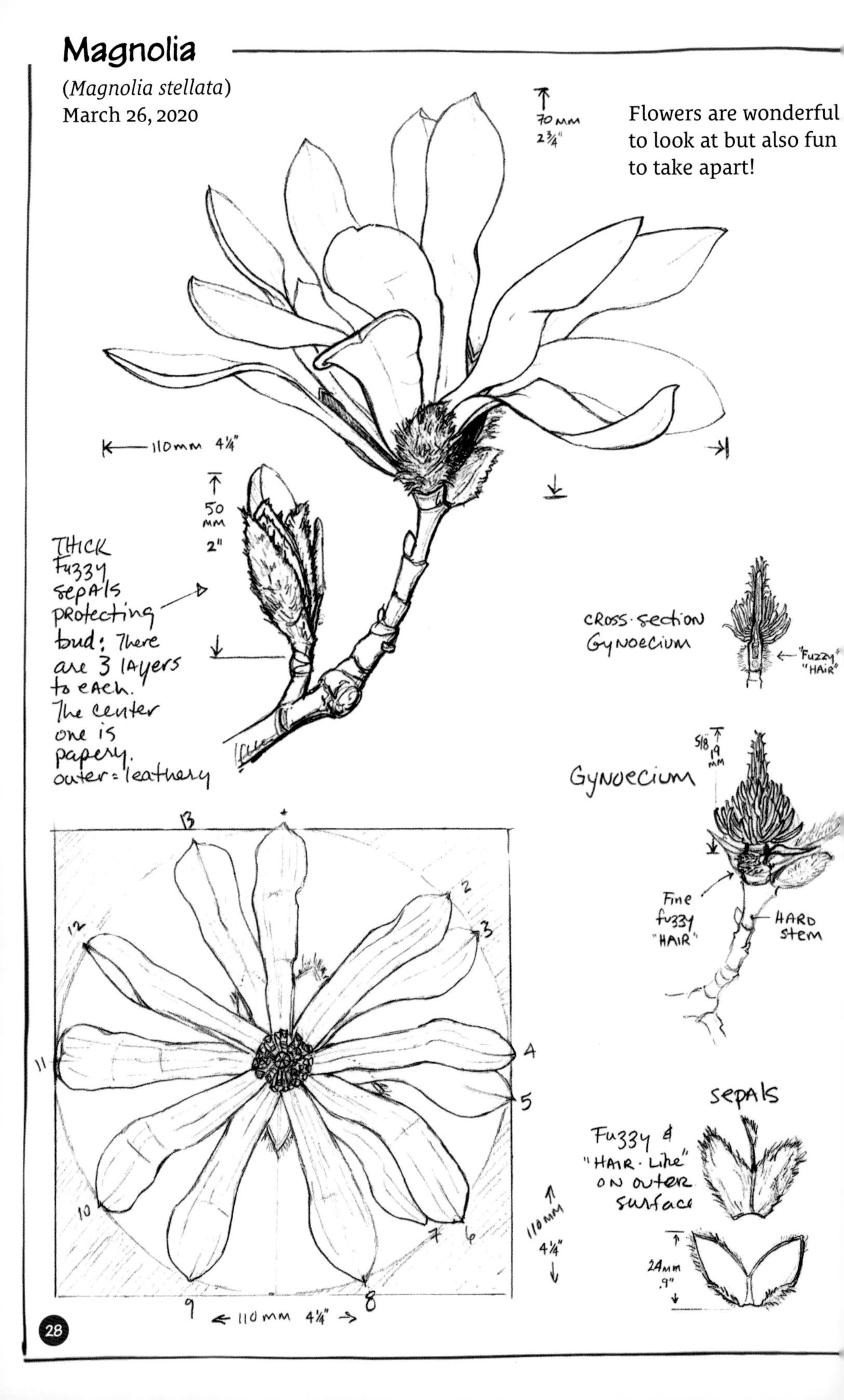
Magnolia
(Magnolia stellata)
March 26, 2020
Flowers are wonderful to look at but also fun to take apart!
70 MM
2¾"
110mm 4¼"
50 MM
2"
THICK Fuzzy sepals protecting bud: There are 3 layers to each. The center one is papery. outer = leathery
CROSS-SECTION Gynoecium
"Fuzzy" "HAIR"
Gynoecium
5/8 19 MM
Fine fuzzy "HAIR"
HARD stem
sepals
Fuzzy & "HAIR-LIKE" ON OUTER SURFACE
24MM .9"
110MM 4¼"
← 110MM 4¼" →
13
2
3
4
5
6
7
8
9
10
11
12

Hornbeam

(*Carpinus betulus*)
October 23, 2021

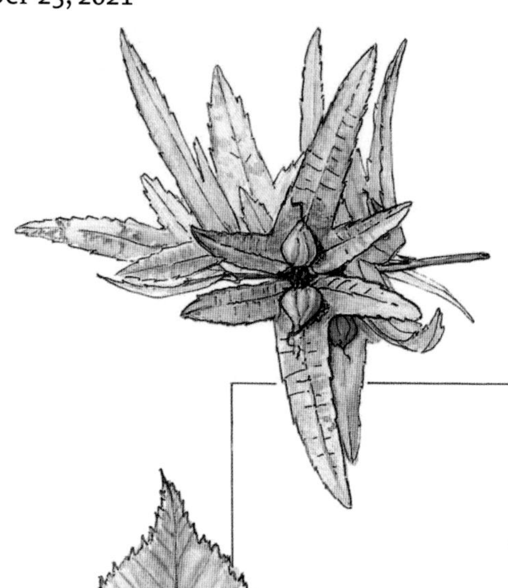

HORNBEAM
(Carpinus betulus)

- Love how the samaras almost always develop in pairs – facing each other like a mirror. At first glance they look like chains of flowers or bells.

- Seeds stay tightly attached to samaras long after they fall from the tree.

- Wood is extremely hard – the hardest of any tree in Europe.

- Traditionally used for ox yokes – the "beam" being attached to their horns.

- deeply furrowed

- wind pollinated.

Oct. 23, 2021
10°C
Sunny breaks

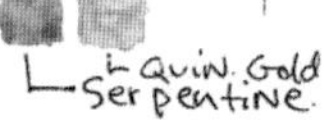

Vancouver Parks

Pacific Spirit Park

Pacific Spirit Regional Park lies within the University Endowment Lands and is located between Vancouver's city limits and the University of British Columbia. This park was established in 1989 by the Metro Vancouver Park system as a natural forest preserve and continues to support a tremendous diversity of plants, animals, bird life, and fungi. The park covers 763 hectares (2.9 square miles), including beaches, marshland, streams, an extensive forest, and 50 kilometres (31 miles) of trails stretching from Burrard Inlet to the Fraser River. Camosun Bog can be found on the eastern border near West 16th Avenue. The map shows one of my favourite sections of the park, north of Southwest Marine Drive and west of Camosun Street.

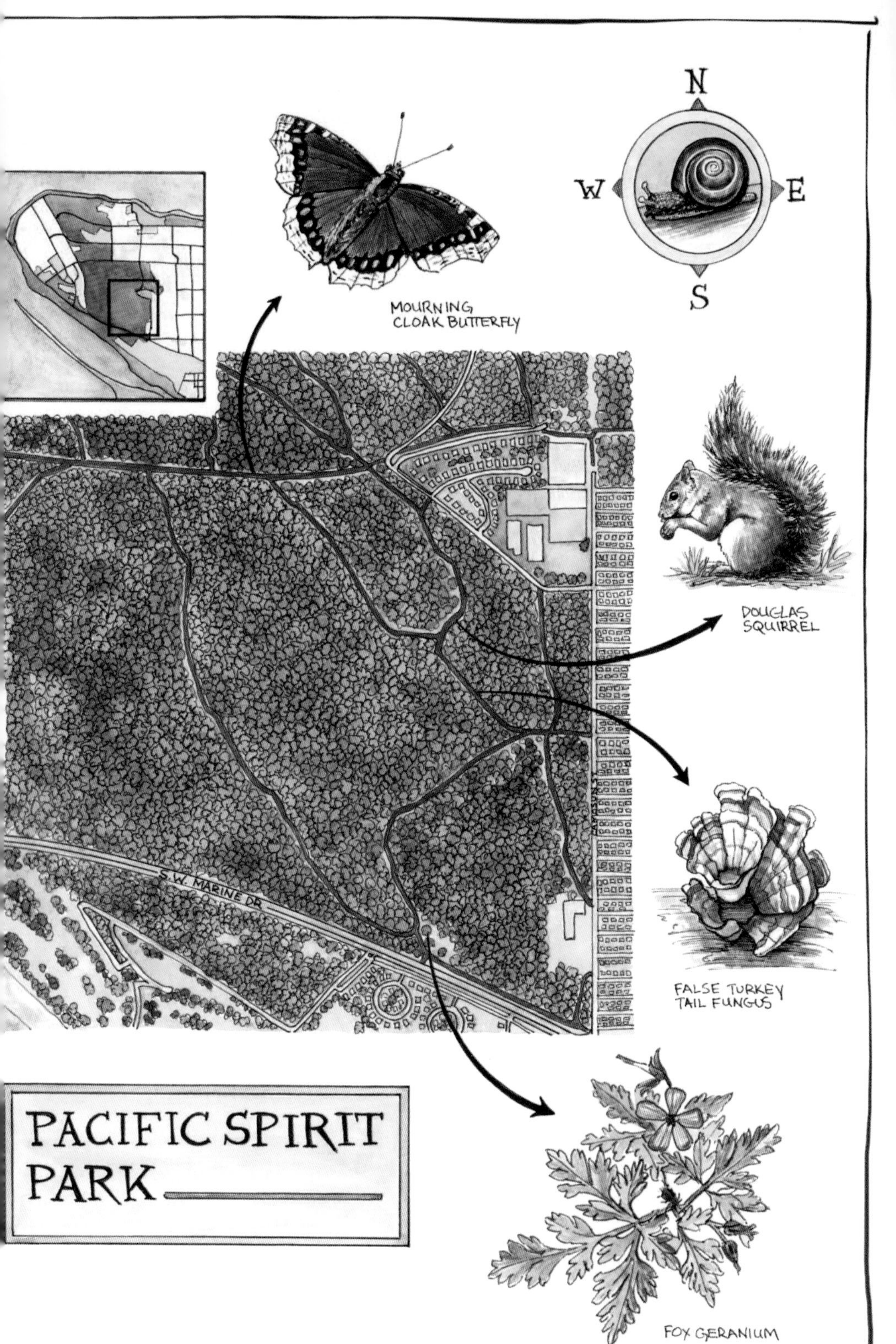
N
W
E
S
MOURNING CLOAK BUTTERFLY
DOUGLAS SQUIRREL
FALSE TURKEY TAIL FUNGUS
FOX GERANIUM
S.W. MARINE DR.
PACIFIC SPIRIT PARK

Great Horned Owl

(*Bubo virginianus*)
April 30, 2020

Like all owls, great horned owls are covered in soft feathers. The leading edge of each flight feather has serrated "combs" that break the air in front of them enabling silent flight. This is an important stealth tactic for hunting and pursuing prey. These owlets, about six weeks old, still looked like balls of light brown fluff high up on a sunny tree branch. They should start to test their wings in about one week.

Owls nest in tree hollows, typically adopting a nest abandoned by other bird species, and lay one clutch of eggs per year. Great horned owls are monogamous and stay in the same territory throughout the year. This owl pair has successfully raised clutches of chicks for the last five years.

Great horned owl talons grow up to 1.3 centimetre (a half-inch) in length and can exert pressure ten times stronger than a typical human grip. Feathered feet may help to protect owls from cold weather and additionally serve to sense contact with prey.

Bird's Nest Fungi

(*Nidula niveotomentosa*)
December 19, 2021

SASAMAT TRAIL

BIRD'S NEST FUNG
DEC 19, 2021
CHILLY! 4°C
ESP. IN THE SHADE
OF THE FOREST
NO WIND

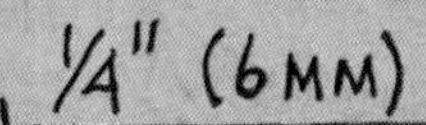

"nest" = peridium = "splash bucket"
"eggs" = peridioles → contain fungal spores

What a story in this tiny world! Bird's nest fungi are roughly the size of your pinky fingernail and have evolved a perfect system to reproduce in damp, shady environments. When it rains, the fruiting bodies or "eggs" get splashed out of the "nest" and can travel more than 1 metre (4 feet). Each one has a sticky tail that wraps around or adheres to twigs or leaves. When the "egg" dries out, it cracks open and spores are released.

Once I learned about these fungi, I searched for years to find them in vain. Finally, on a sunny day in December – success! Incredibly, I had walked right by them more than a dozen times. All lined up in tidy rows on a wooden fence railing, they waited patiently to be noticed.

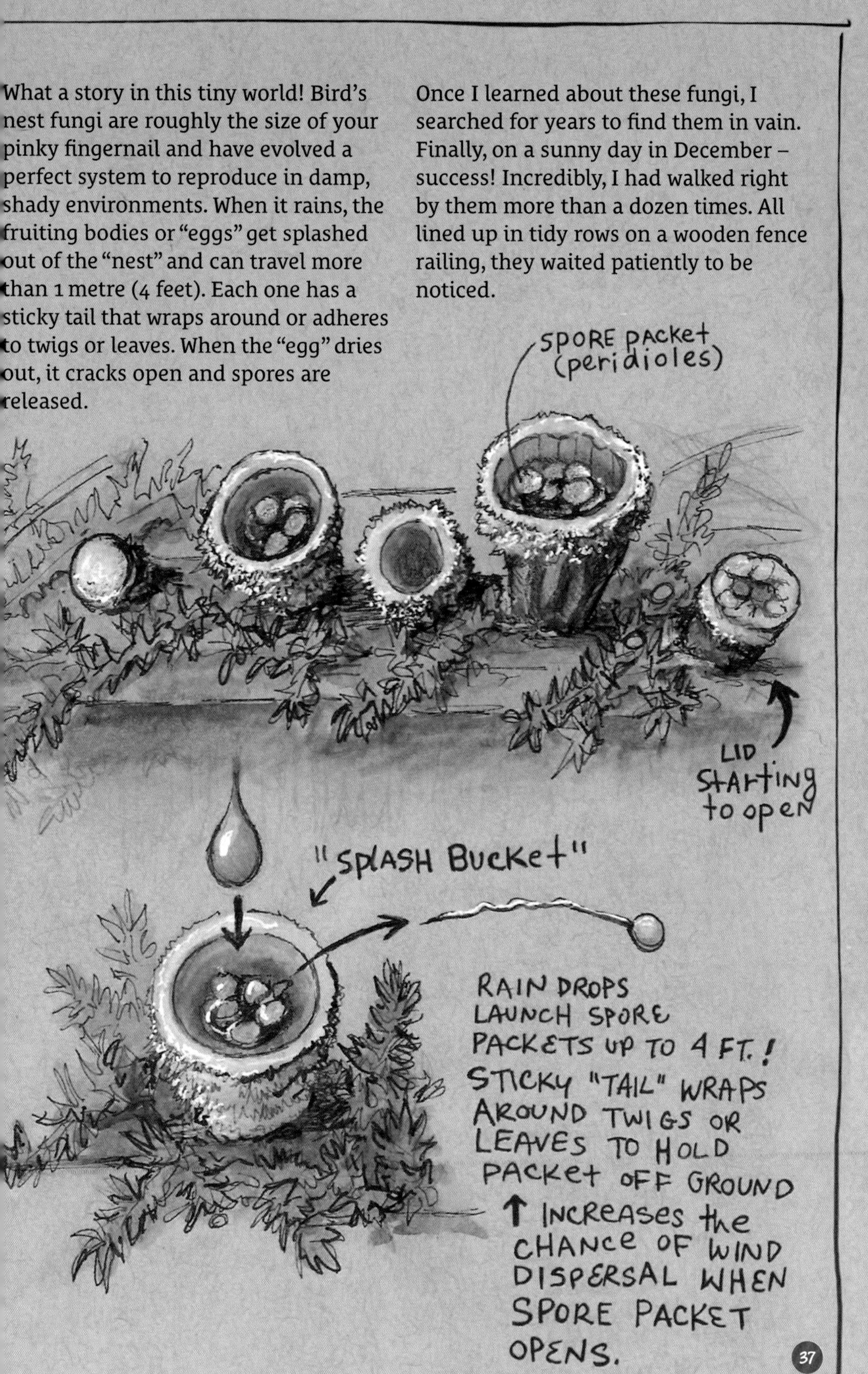

Slime Mold

(*Physarum polycephalum*)

January 20, 2022

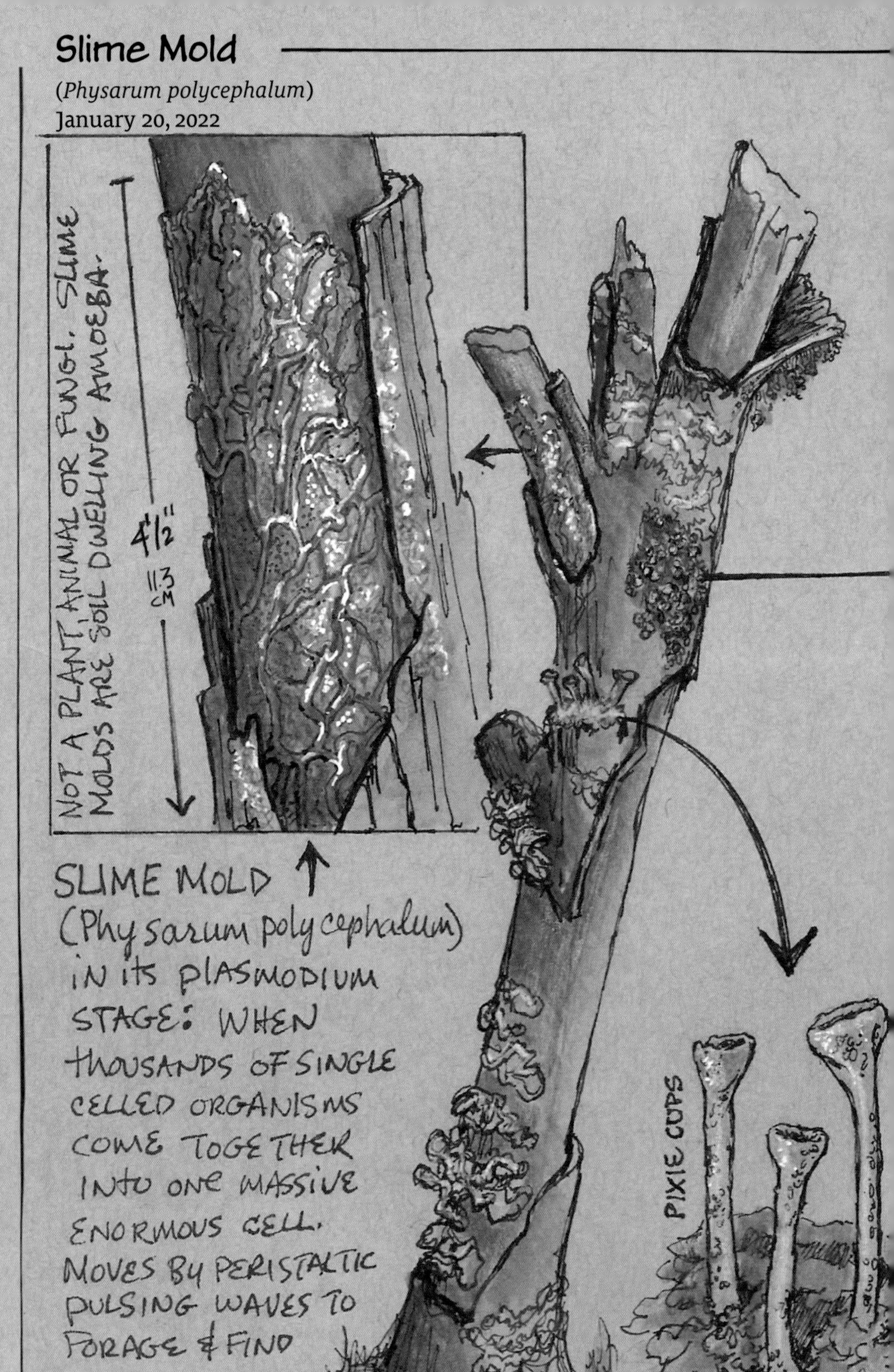

Slime molds are not actually molds but are similar to single-celled amoebas. Also referred to as myxomycetes, plasmodial slime molds can be found on decaying forest litter and rotting wood when conditions are right. They are important agents that decompose and recycle nutrients in the food web.

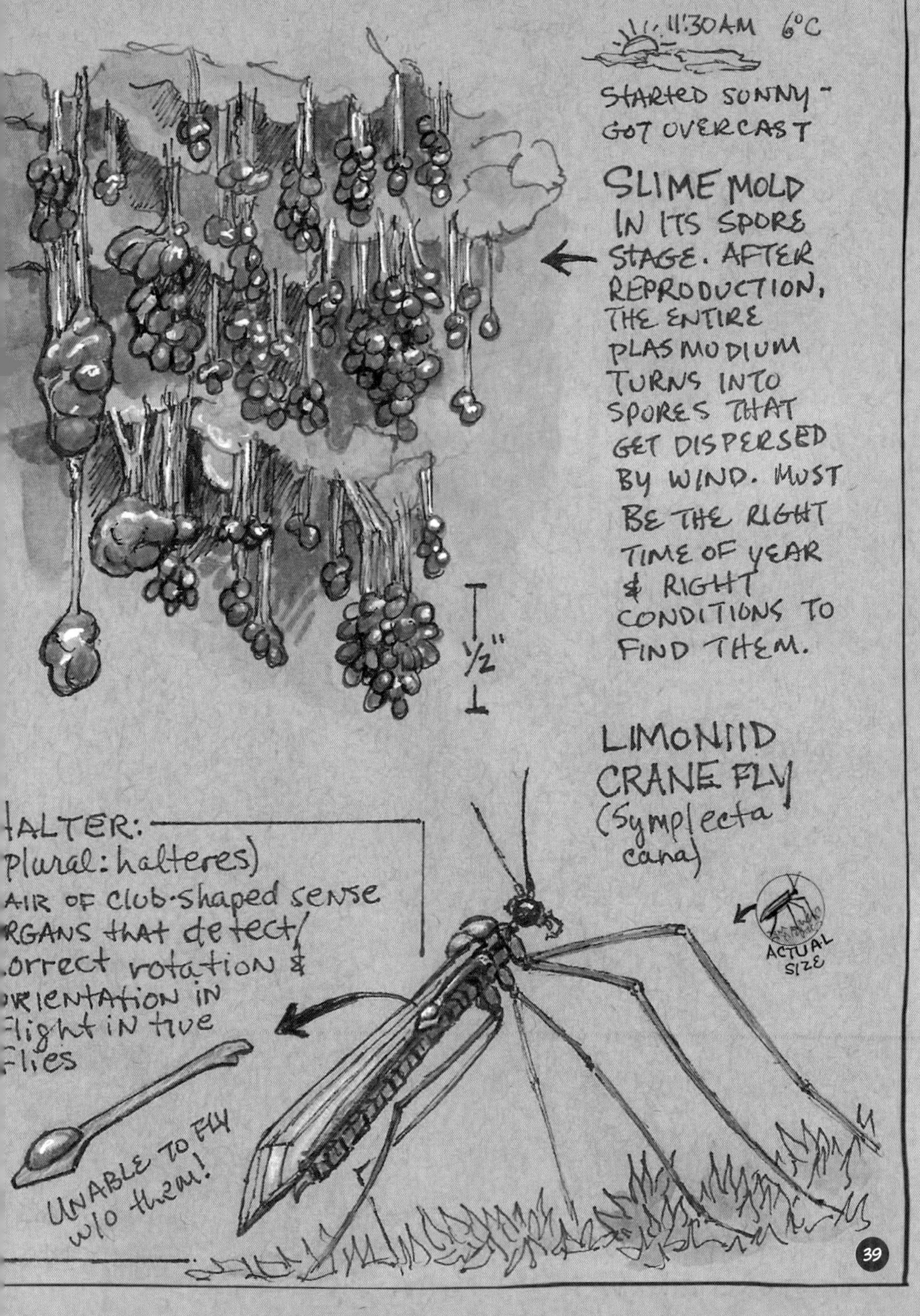

A Forest Walk

January 20, 2022

Shaggy Scalycap

(*Pholiota squarrosa*)
September 16, 2021

(Pholiota squarrosa) → "Squarros = "upright scales, rough, scrubby"

Camosun Bog

The Camosun Bog took shape five thousand years ago when ice carved out a depression in the land to create a lake. This lake turned into a swamp, which then turned into a peatland bog. Bogs are important because they store large amounts of carbon. Native plants that thrive in low oxygen, acidic environments, such as the keystone species sphagnum moss, are essential in any bog. Sphagnum moss manipulates its habitat by absorbing water that it then releases back into the environment during dry conditions. Sphagnum turns to peat when it decays and dies as it sinks to become the bottom layers of the bog. When housing developments began to increase in the area, the City of Vancouver put in drains to eliminate excess groundwater. In essence, this dried out the bog and allowed large trees like hemlock to encroach and begin to take over the area. The Camosun Bog Restoration Group was established in 1995 to restore Camosun Bog with native bog plants like sphagnum moss, sundew and labrador tea. Self-dubbed the "Crazy Boggers," this group continues to maintain and care for this vital ecosystem in the Pacific Spirit Regional Park.

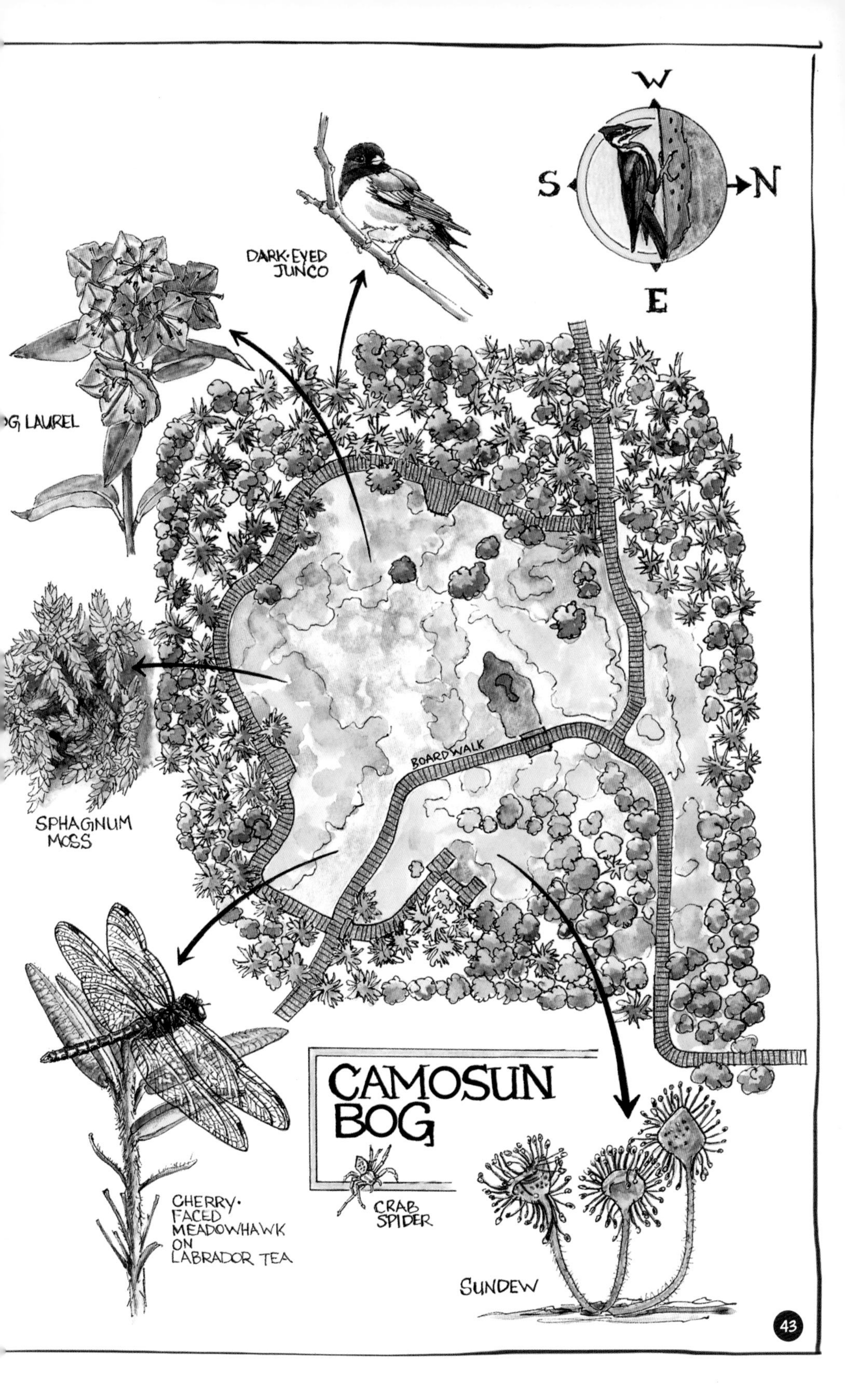
W
S
N
E
DARK-EYED JUNCO
OG LAUREL
SPHAGNUM MOSS
BOARDWALK
CAMOSUN BOG
CHERRY-FACED MEADOWHAWK ON LABRADOR TEA
CRAB SPIDER
SUNDEW

November 13, 2021

LICHEN
= "CORALS OF THE FOREST"
CAMOSUN BOG NOV. 13, 2021
• VERY sensitive to air pollution (So not often seen in cities)

Lichens are actually 3 organisms: functioning as 1 single stable unit:
① Basidiomycete fungus
② Ascomycete fungus
③ Green alga
= mutually beneficial relationship

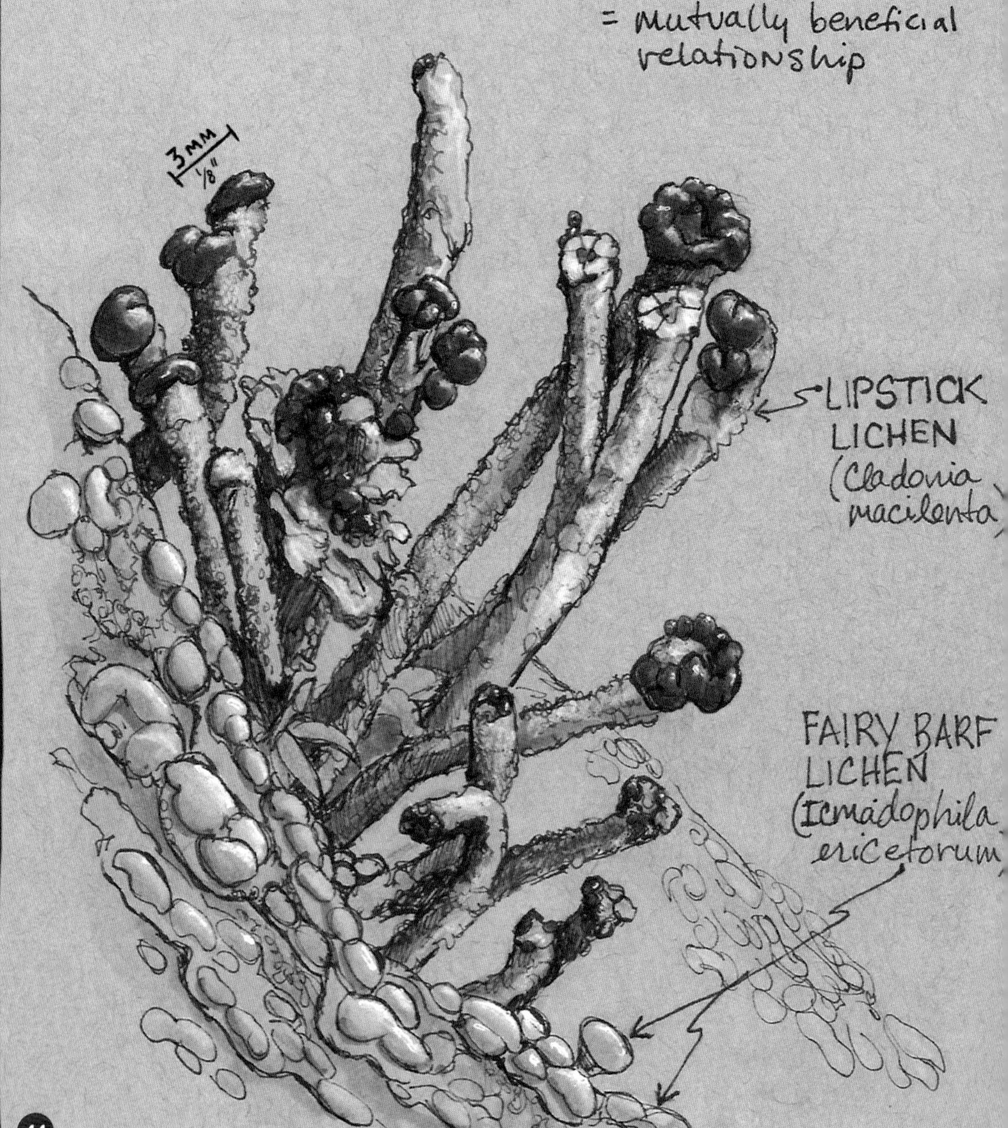

Surprisingly, with over 3,600 species of lichens in North America and 1,600 species of fungi in British Columbia, it is still easy to pass them by! Many lichen structures and colours resemble corals found in the ocean. Lichens are important in recycling nitrogen, a key nutrient for the growth of other plants in the ecosystem, and help stabilize soils from erosion. Due to increasing air pollution, lichen populations have been decreasing globally, but they are still thriving in the Pacific Northwest.

Actual cup height 1/4" 6mm

MYCENA SP.

HAIRY CURTAIN CRUST AKA. FALSE TURKEY TAIL FUNGUS (stereum hirsutum)

ENDOMENT LANDS
CAMOSUN TRAIL
overcast
7°C NOV 13, 2021

Douglas Squirrels

(*Tamiasciurus douglasii*)
July 24, 2020

Contemplating how to capture the 20-centimetre (7.5 inches) thick gnarly bark of a Douglas fir on paper, my thoughts are punctured by a clatter of claws behind me. I wheel around and spot the culprit, but not before being pelted with bright-green pinecone scales. This tiny Douglas squirrel makes quick work of her namesake pinecone. With efficient grace, she peels off individual scales of the cone in sequence. Holding each scale like a dish, she scrapes out the two-seed treasure with her teeth and gobbles them down before flinging the scale away.

I just happen to be standing in her family's "midden" – the jumble of discarded "dishes" that pile up under favourite squirrel dining spots. The core of the cone is held tightly under one arm as this tiny squirrel spirals her way down to the tip, eating the seeds as she goes.

Angel Wings

(*Pleurocybella porrigens*)
December 5, 2020

On a dreary December day, it was hard to miss this beautiful bright-white mushroom along the edge of the trail. Angel wing mushrooms look very similar to oyster mushrooms in shape, but oysters are rarely all white in colour.

ALONG IMPERIAL TRAIL
- Decomposes wood "porrigens" means "spreading"
- Gills run the full length of under side
- Almost always grows on conifer wood - often hemlock
- Spore print is white

BRIGHT WHITE

Controversity re: toxicity!

3:45pm
Dec. 5, 2020

- Don't have a stalk

10°C

Brackets

Polypores (*Fomes fomentarius*)
February 10, 2021

I love looking closely at bracket fungi. They may appear dullish grey from far away, but the surface of each cap usually has intricate lines, patterns, and shapes – almost like a long-lost alphabet or code.

Brackets are a type of fungi that grow on dead or dying wood. Their spores are released through tiny pores on the underside of their caps instead of through gills like other mushrooms.

Common names include tinder polypore and hoof fungus because brackets are useful as tinder to start campfires and are U-shaped, resembling a horse's hoof. The scientific name *Fomes fomentarius* translates to "tinder used for tinder."

Also called conks, this tinder bracket has been used by humans for at least five thousand years. The "Iceman," or Ötzi – the oldest human mummy found in the Ötztal Alps of Italy – had with him a fire-starting kit that included tinder polypore. It's quite incredible to make a connection to Neolithic times by simply taking a walk through a Vancouver forest!

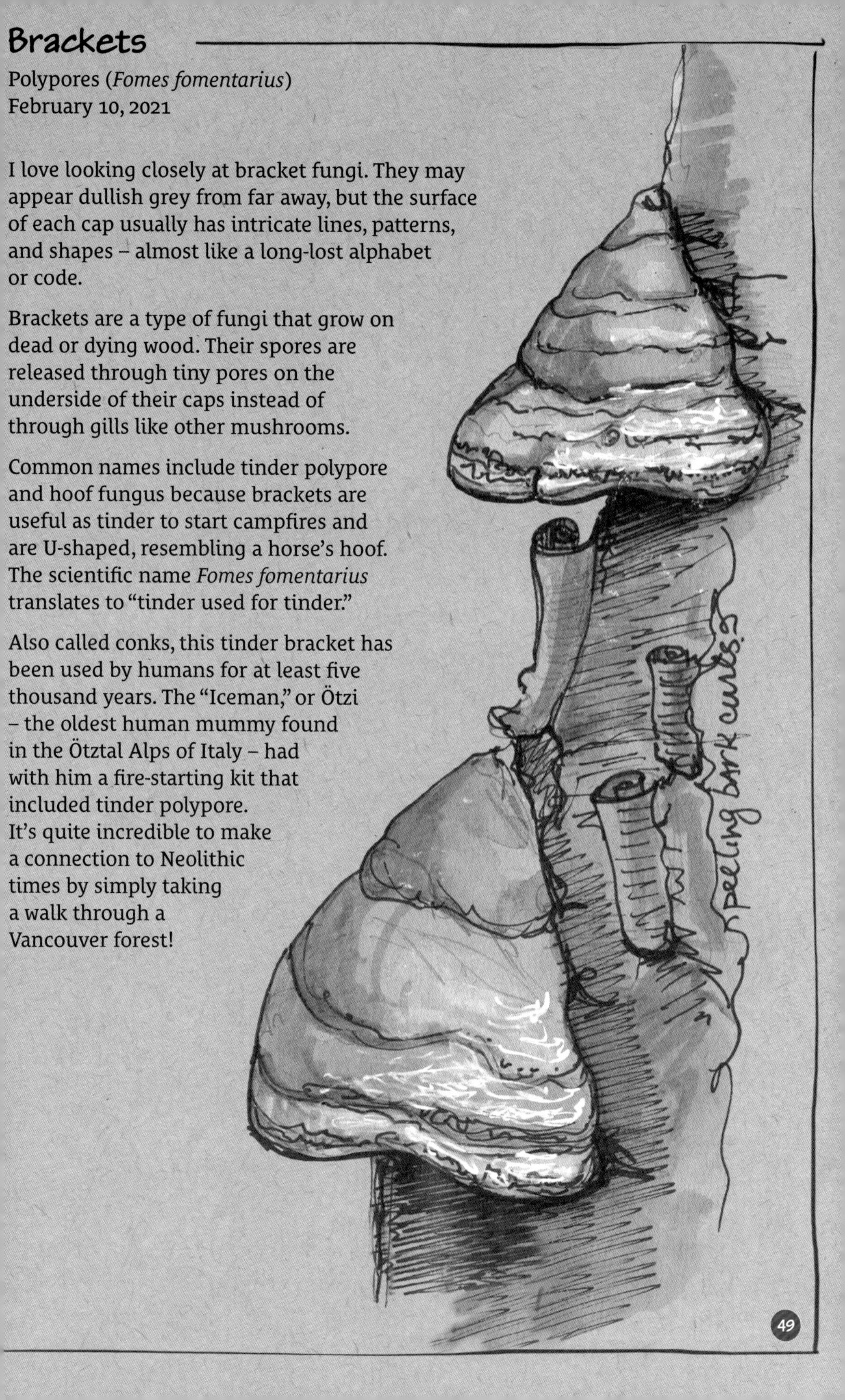

Pileated Woodpecker

(*Dryocopus pileatus*)

June 29, 2021

Western Polished Lady Beetle

(*Cycloneda polita*)
January 20, 2022

3:30 PM
5°C

- THOUGHT THIS WAS A "BABY" LADY BUG, BUT NO: SPOTS DO NOT DESIGNATE AGE – THEY DESIGNATE SPECIES

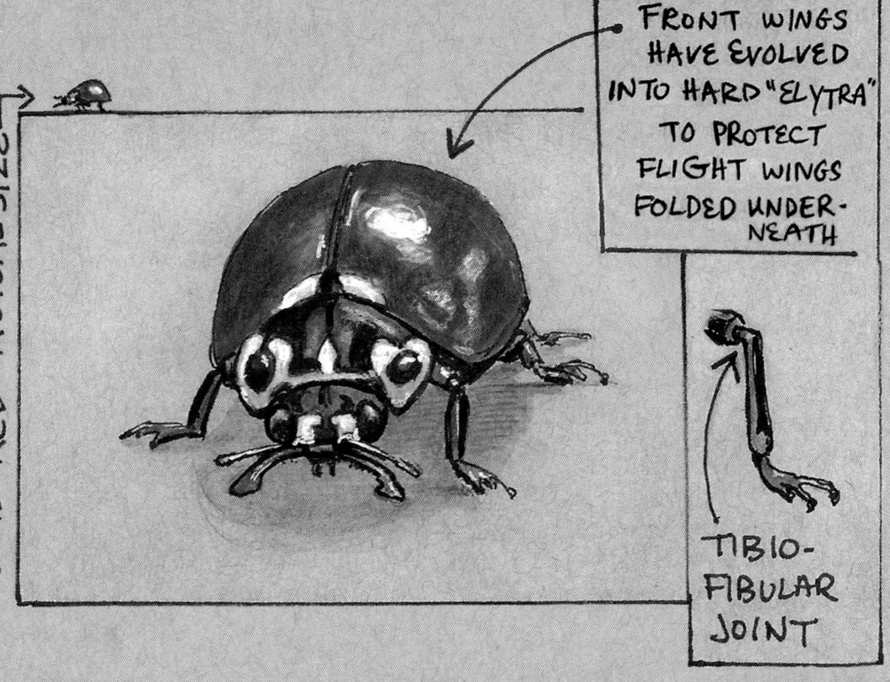

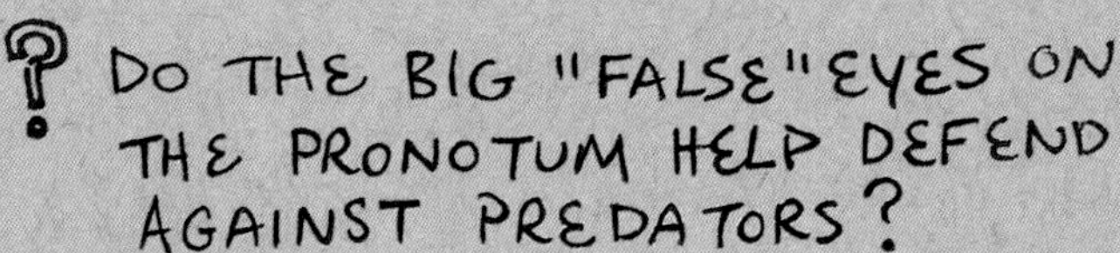

? DO THE BIG "FALSE" EYES ON THE PRONOTUM HELP DEFEND AGAINST PREDATORS?

- BRIGHT RED COLOUR IS A MAIN DEFENSE = APOSEMATIC COLORATION: SIGNALS IT IS TOXIC TO PREDATORS.
- LADY BEETLES SECRETE FOUL SMELLING DROPLETS OF ALKALOID CHEMICALS FROM TIBIO FIBULAR JOINTS = HEMOLYMPH.

CALLED "REFLEX BLEEDING" WHEN THREATENED

Jericho Beach Park

Jericho Beach Park was once the location of the ancient Musqueam Nation village Ee'yullmough. The beach here became the site of the Jerry & Co. logging operation owned and operated in the 1860s by Jeremiah Rogers. The name "Jericho" is believed to have come from the name "Jerry & Co." or the area's nickname "Jerry's Cove." A large golf course created by the Jericho Country Club replaced the logging company between 1890 and 1919. In the 1920s, the Royal Canadian Air Force built army barracks, flying boat hangars, and a military airstrip, which took on a defensive role during WWII. Following a public campaign in 1973, the land was sold back to the City of Vancouver and designated as parkland. Park landscaping was completed in 1989. The Jericho Stewardship Group, composed of residents, students, naturalists, and nonprofit groups, continues to work with the Vancouver Board of Parks and Recreation to restore and enhance this habitat.

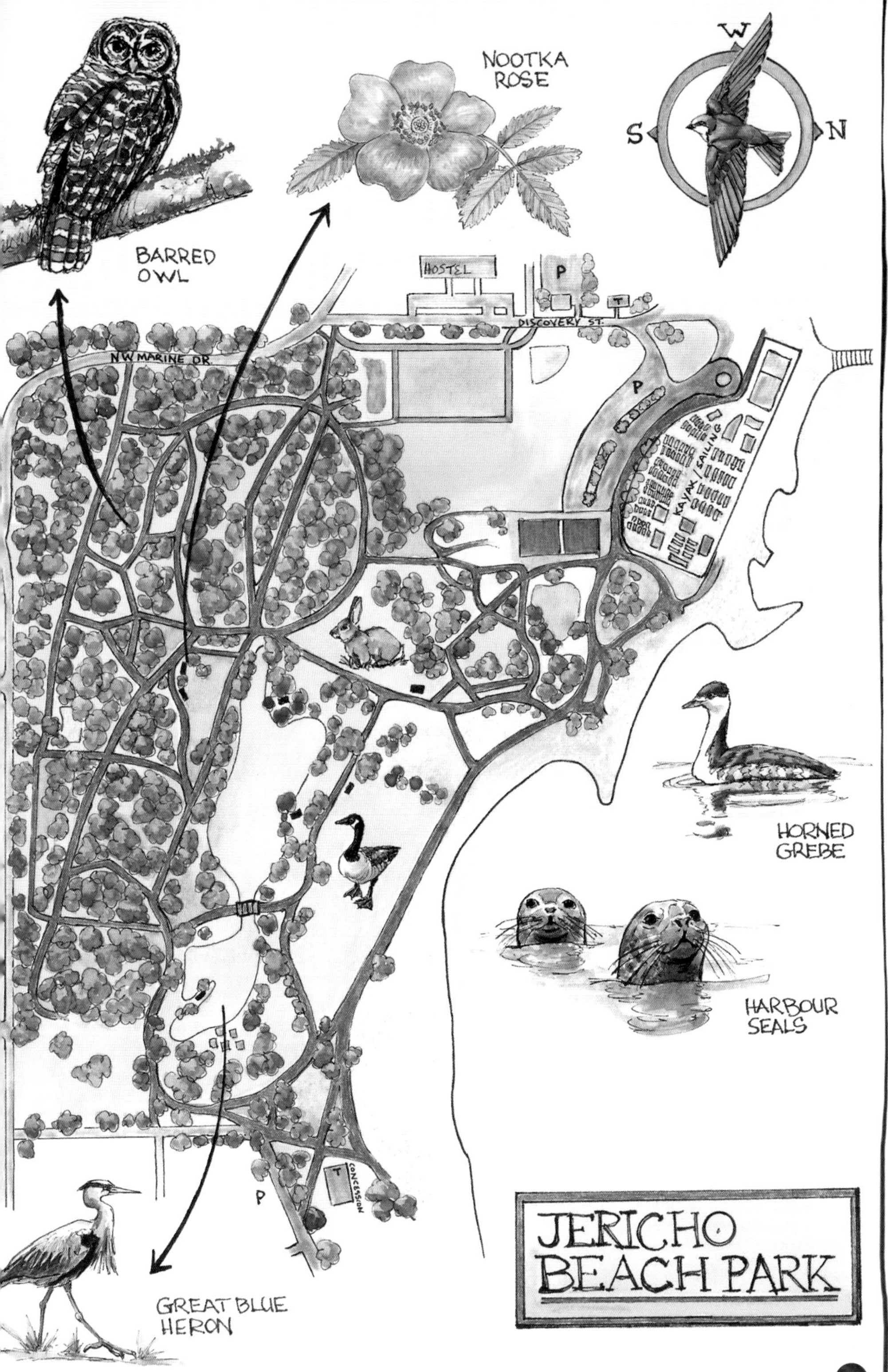

JERICHO BEACH PARK

Great Blue Heron

(*Ardea herodias*)
October 23, 2021

Jericho Stream
December 11, 2021
JUST BEFORE
THE RAIN

Black Poplar Leaves

(*Populus nigra*)
October 21, 2021

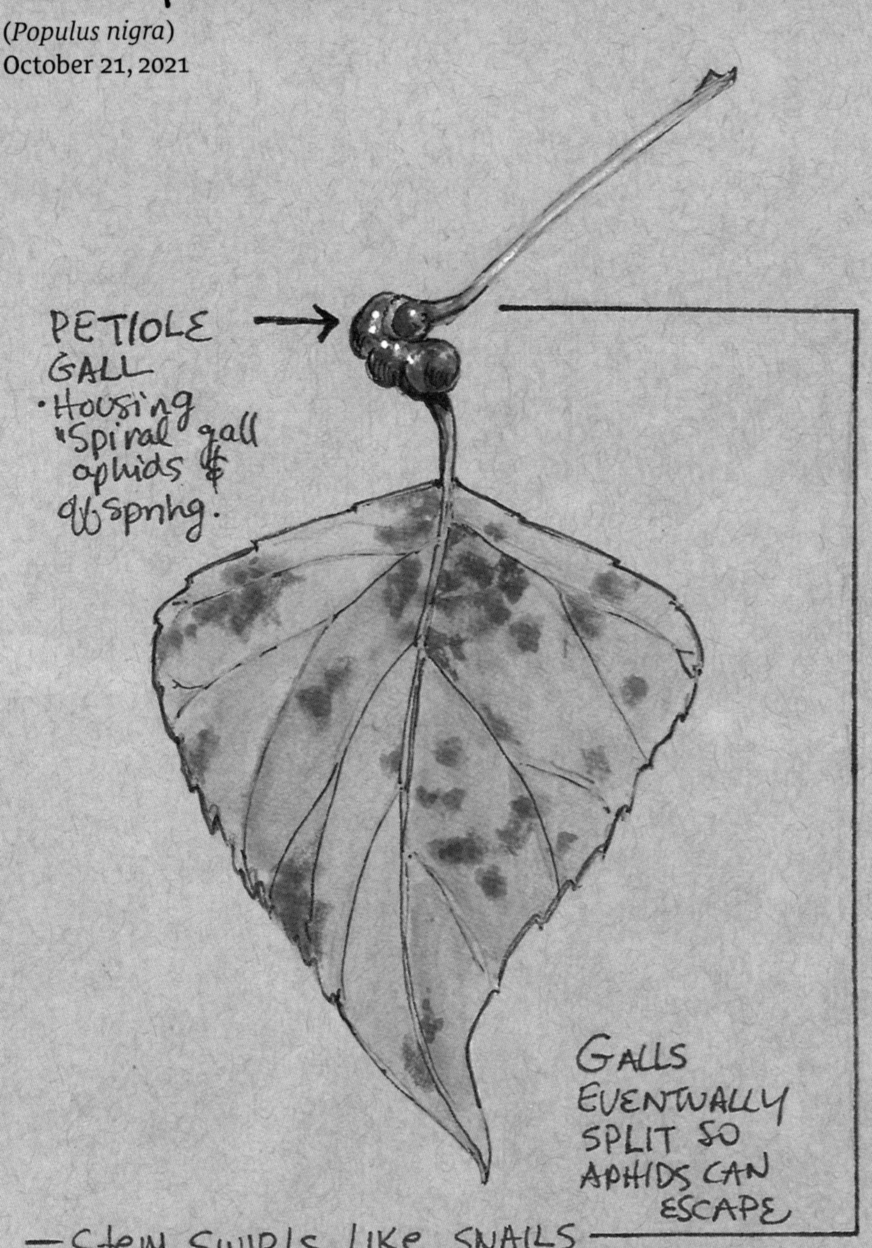

— Stem swirls like snails uninvited, or lost marbles by children in the park.

They are petiole galls! → Relatively harmless to tree.

(Petiole = stem) made by aphids (*Pemphigus spyrothecae*) chewing on stem

JERICHO
OCT 21, 2021
2 PM WARM!
18°C RAIN COMING!!
BLACK POPLAR
(Populus nigra)

Horsetail

(*Equisetum telmateia*)
April 8, 2021

Seeing horsetail always makes me think of dinosaurs! It's quite incredible (and encouraging) to have plant species from 270 million years ago still thriving in our city. The visual patterns created by horsetail are mesmerizing to draw. If you spot them on your outing, take a moment to observe and draw these prehistoric plants.

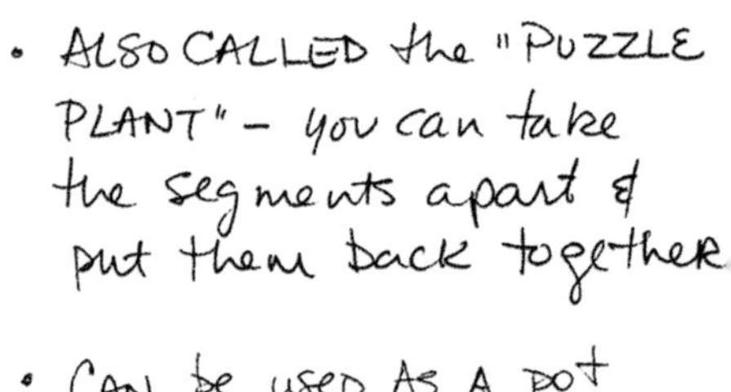

- CAN be used AS A pot scrubber while CAMPING. Contains A high silicA content that MAKES it abrAsive. CAN ALSO BE USED AS SANDPAPER
- Its Roots are aggressive and important to bind soils of banks & shoreline preventing erosion.
- THEY REPRODUCE BY SPORES LOCATED IN THE TOP "CONE". & IS RELATED TO FERNS.

Barred Owl

(*Strix varia*)
May 22, 2021

I always pay attention when I hear a group of birds making a ruckus. In this case, it was a flock of crows. They were unhappy that this barred owl decided to nap in the vicinity of their nest – or perhaps fledglings were nearby. Incredibly, the owls never seem overly bothered by the barrage of irate yells and fluttering crow wings inches from their heads.

Unfortunately, barred owls outcompete the endangered northern spotted owl for habitat and food. Other threats to spotted owls include habitat loss and fragmentation of old-growth forests. There are only six spotted owls left living in the wild in all of Canada. The mission of the Northern Spotted Owl Breeding Program in Langley, a municipality located in the eastern part of Metro Vancouver, is to breed spotted owls in captivity and release them into protected habitat to prevent this species from becoming extirpated across Canada.

Leucistic Fox Sparrow

(*Passerella iliaca*)
March 24, 2022

First noticing a small group of songbirds foraging on the ground by a hedge of Nootka roses, I spotted this white ball of fluff against the muddy underbrush. All I could think of was a bouncing snowball! But what kind of bird was it? Luckily, the triangles on its chest were fairly visible. I was excited to realize this was a leucistic fox sparrow.

- Leucism is a condition causing a partial loss of pigmentation in animals and birds, resulting in white, pale, or patchy coloured skin or feathers.
- Some feathers in birds with leucism lack melanocytes – the cells that produce the pigment melanin.
- Albinism, in comparison, is a rare group of genetic disorders that cause the skin, hair, and eyes to have very little or no colour.

Common Fox Sparrow

(*Passerella iliaca*)
March 24, 2022

(Passerella iliaca)
↓ "little sparrow"
→ "flat/triangular"

- The fox sparrow forages using the "double scratch" technique: it takes one hop forward, then immediately scratches the soil and leaf litter backwards.
- Adults are known to perform a "broken wing" display to draw predators away from their nest or young.
- Fox sparrows look very similar to song sparrows. How to tell the difference? Look for the triangles on their chest and flanks compared to the vertical streaks on a song sparrow.
- Four different groups of fox sparrows and eighteen subspecies can be identified based on plumage, migration, and song.

∧ MARKS = chevrons

☆ Note! yellow on bill turns pinkish during breeding season

Fraser River Park

In 1995, development pressure along the Fraser River raised public interest to conserve the large, undyked Fraser River floodplains. This led to acquisition of the land by BC Parks. Located along the mighty Fraser River, Fraser River Park opened to the public in June 2000. It includes walking trails, wooden boardwalks and bridges, picnic tables and benches among stands of poplar, alder, and crab-apple trees. Native Nootka roses also line many of the pathways.

Just west of the park, the most recent addition to Vancouver's only riverside greenway system is the Fraser River Trail which opened in 2010. It follows tightly along the north shore of the North Arm of the Fraser River and, when complete, will connect Pacific Spirit Regional Park to the City of Burnaby's trail system. Currently, McCleery Golf Course, Fraser River Park, Shaughnessy Street Park, Gladstone Park, and Riverfront Park waterfront walkways and bikeways are part of this greenway.

Deering Island Park is located along the Fraser River trail on a small island between the Point Grey golf course and the Celtic Shipyards.

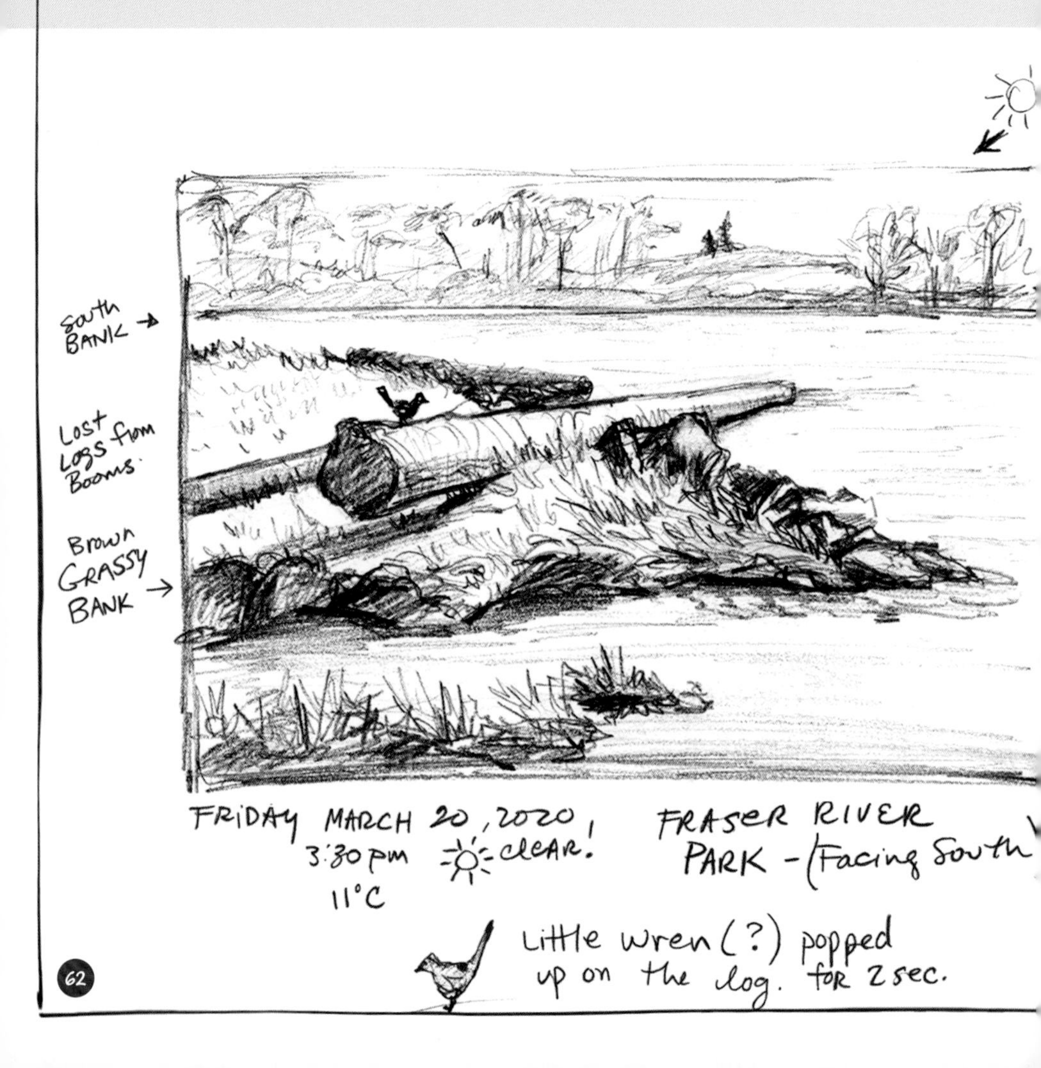

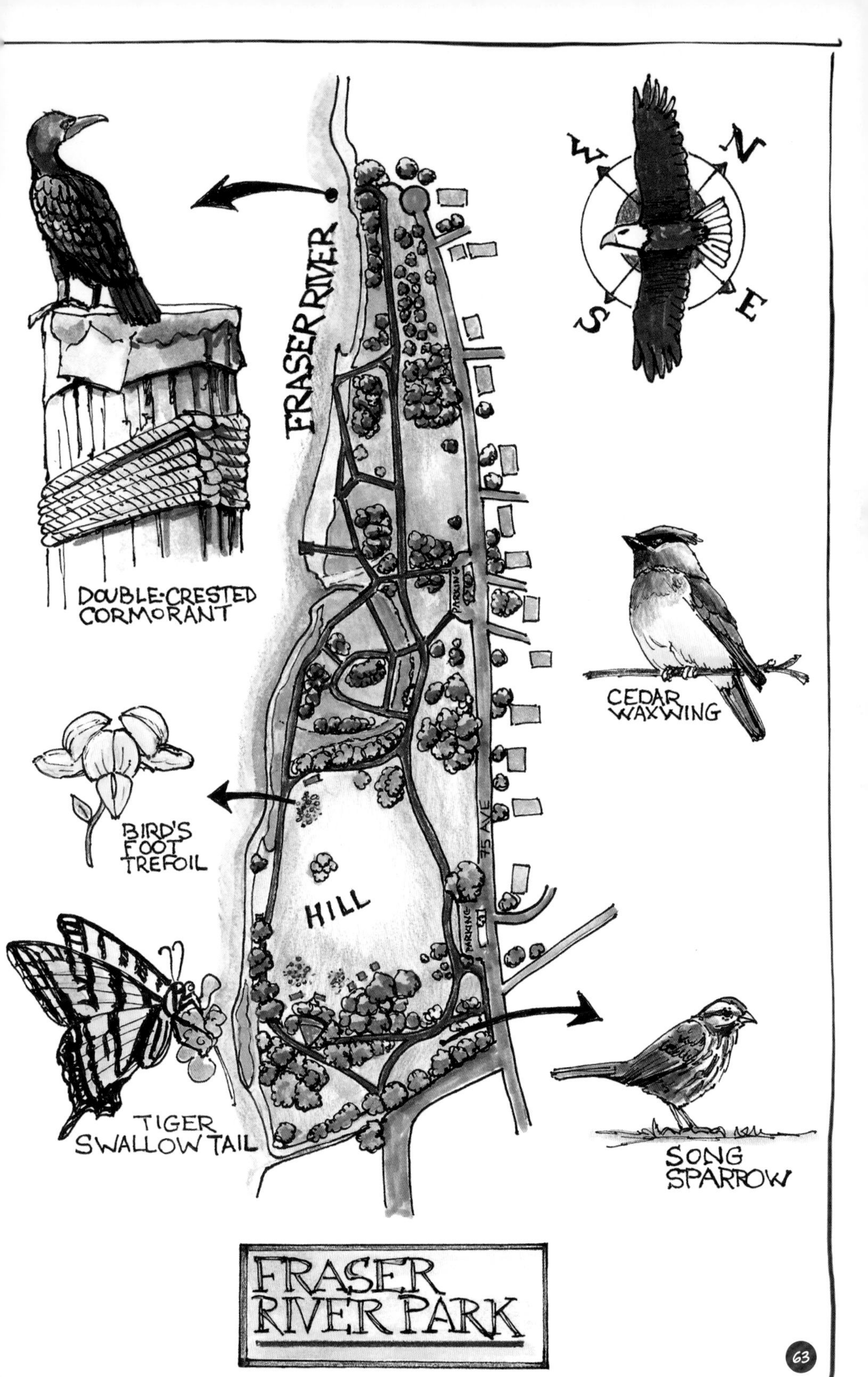
FRASER RIVER
W
N
S
E
DOUBLE-CRESTED
CORMORANT
CEDAR
WAXWING
BIRD'S
FOOT
TREFOIL
HILL
PARKING
75 AVE
PARKING
TIGER
SWALLOW TAIL
SONG
SPARROW
FRASER
RIVER PARK

A Day at Fraser River Park

June 11, 2021

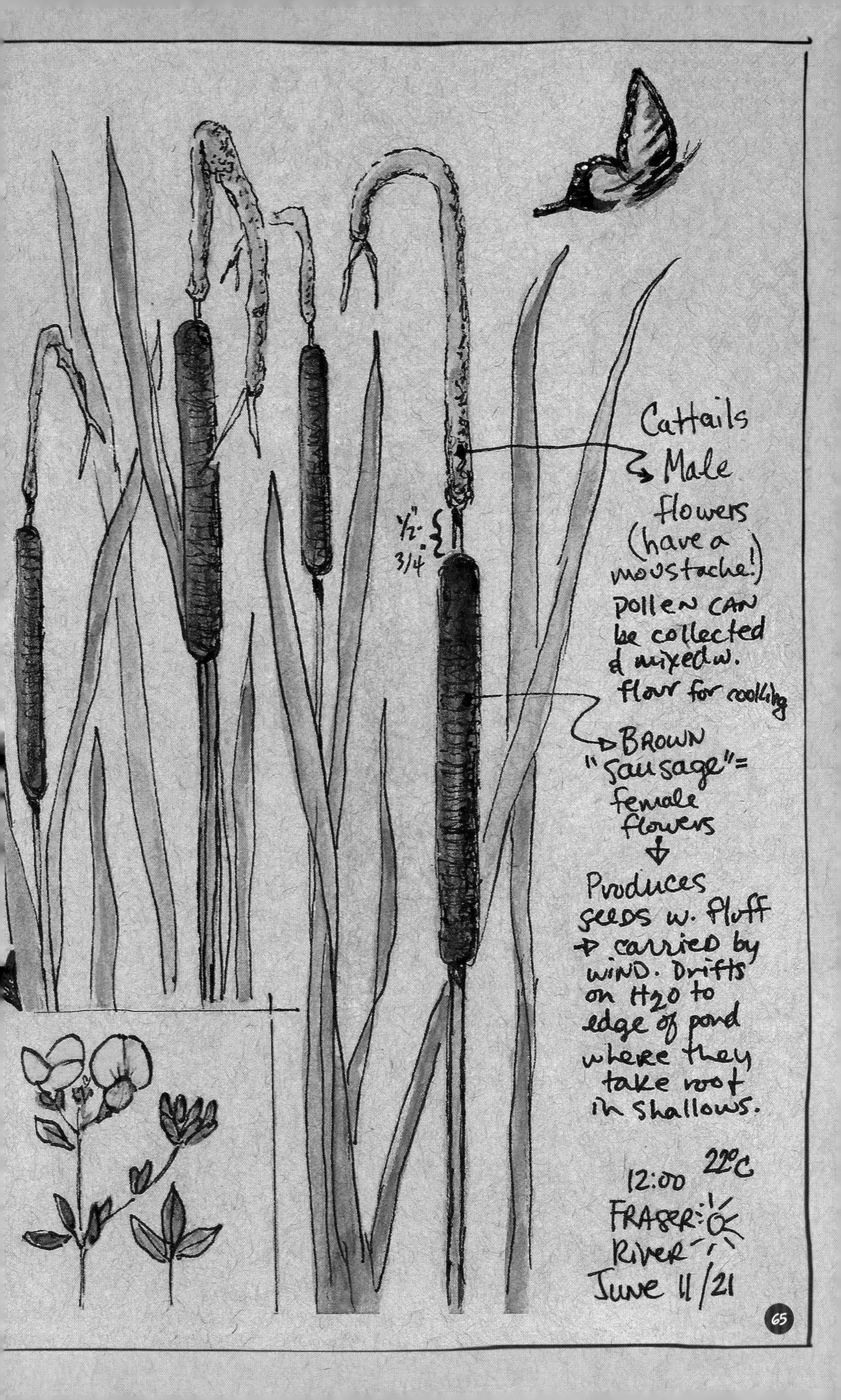
Cattails
Male
flowers
(have a
moustache!)
pollen CAN
be collected
& mixed w.
flour for cooking
BROWN
"sausage" =
female
flowers
Produces
seeds w. fluff
carried by
wind. Drifts
on H_2O to
edge of pond
where they
take root
in shallows.
1/2"-
3/4
12:00
22°C
FRASER
River
June 11/21
65

Just West at Deering Island Park

January 16, 2022

Chicory

(*Cichorium intybus*)
October 11, 2021

Dabbler Ducks

(*Anas*)
February 17, 2021

Dabblers are ducks that feed mainly on the surface of the water or mud flats using a skimming or sweeping motion with their heads. Ducks in this group include mallards, shovelers, teals, pintails and wood ducks. Ducks don't chew their food. Their bills have evolved special features that help them manipulate and position food before swallowing it whole. Comblike structures called lamellae resemble serrated teeth that line the inside edge of their top and bottom bills. These "teeth" filter and strain invertebrates, seeds and plant matter from mud and water as they sweep their heads from side to side while foraging.

Look closely at the upper bill of the wood duck shown below and you can see a small bump on the tip called a nail. Nails are useful for digging through mud or debris and help ducks uncover small roots, seeds, worms and other foods. This male posed perfectly for me and had so much charisma, I couldn't wait to draw him in this studio piece.

The tide was low along the Fraser River when the sparkling water caught my eye. Then I noticed a dozen ducks moving slowly, heads bent to the surface. I had never seen ducks walking along with their bills on the surface of ankle-deep water, moving their heads steadily back and forth, left to right and back again.

McDonald Beach Park

April 29, 2021

Across the Fraser River to the south is McDonald Beach Park in Richmond. This bald eagle looked a bit disgruntled, biding its time while waiting for soggy wings to dry. She likely got soaked while hunting prey in the river. Feathers are extremely heavy when wet, making it difficult to fly. Hopefully the fishing was good!

SEED POD
SCOTCH BROOM
salmonberry

Quilchena Park

Quilchena Park is located along the Arbutus Greenway at West 33rd Avenue. In 1925, this low-lying land was at the edge of Vancouver's city limits. At that time, it was turned into a popular nine-hole golf course that stretched from King Edward Avenue to 33rd Avenue. When the golf course moved to Richmond in 1956, the area was transformed into a park with rolling lawns that form a natural amphitheatre. Today, it remains as a large open area surrounded by poplar, pine, elm, and magnolia trees, making this a perfect spot for picnics, sports get-togethers, and sketching nature's stories. Other public amenities include a walking and jogging trail, a playground, a skateboard park, baseball diamonds, a disc golf course and a small Hellenic garden at the northwest corner.

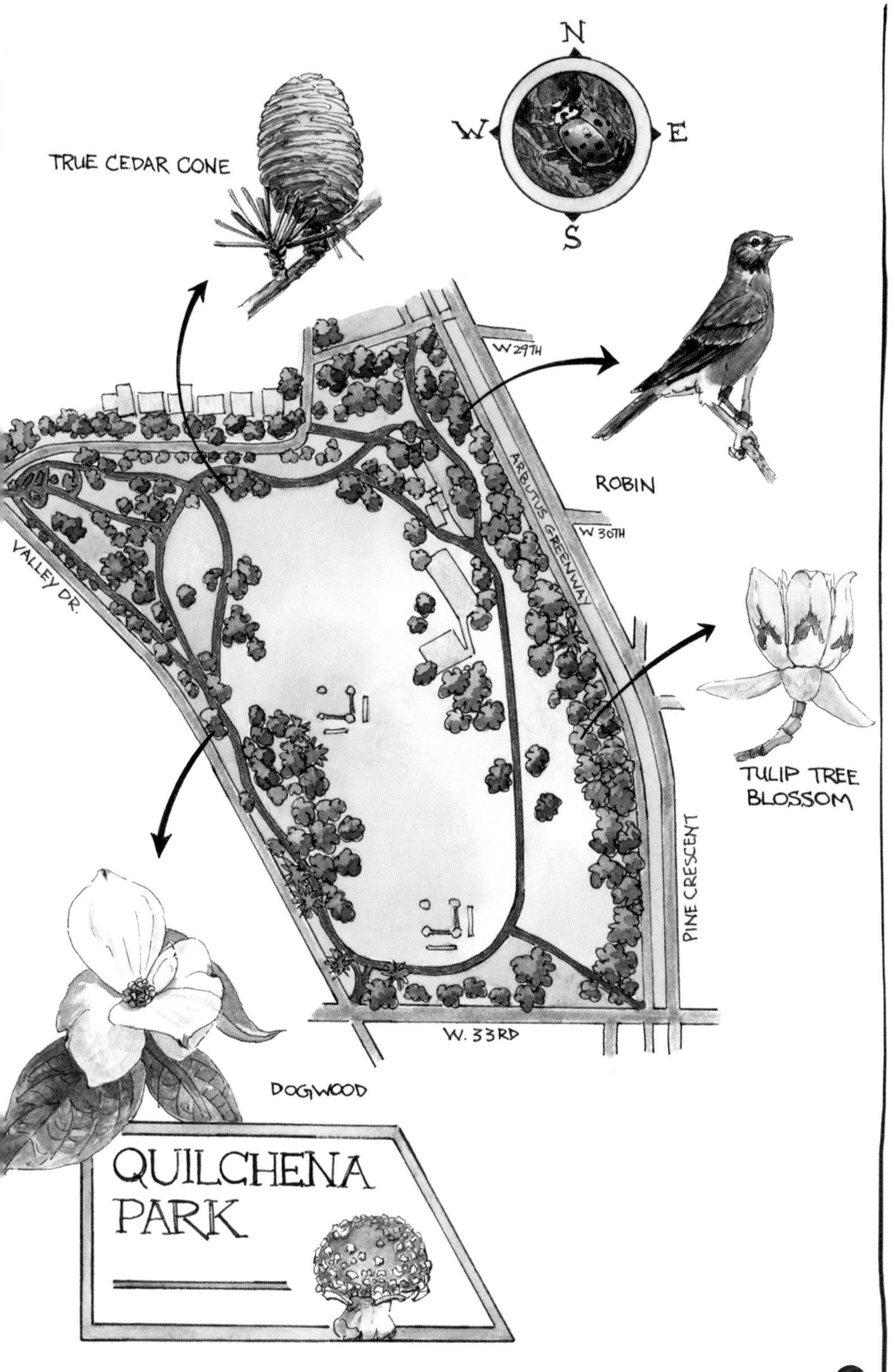
N
W
E
S
TRUE CEDAR CONE
W 29TH
ROBIN
W 30TH
ARBUTUS GREENWAY
VALLEY DR.
TULIP TREE
BLOSSOM
PINE CRESCENT
W. 33RD
DOGWOOD
QUILCHENA
PARK

Tulip Tree

(*Liriodendron tulipifera*)
June 28, 2021

This lovely tree in the magnolia family earned its name for its tulip-like blossoms. The tree's unique seed pods caught my eye. Shortly afterward, I spotted the black and white "storm trooper" hornets buzzing around them. These bald-faced hornets are not hornets at all, but are wasps related to yellow jackets in the family Vespidae.

The tulip tree is a host plant for tiger and spicebush swallowtail butterflies.

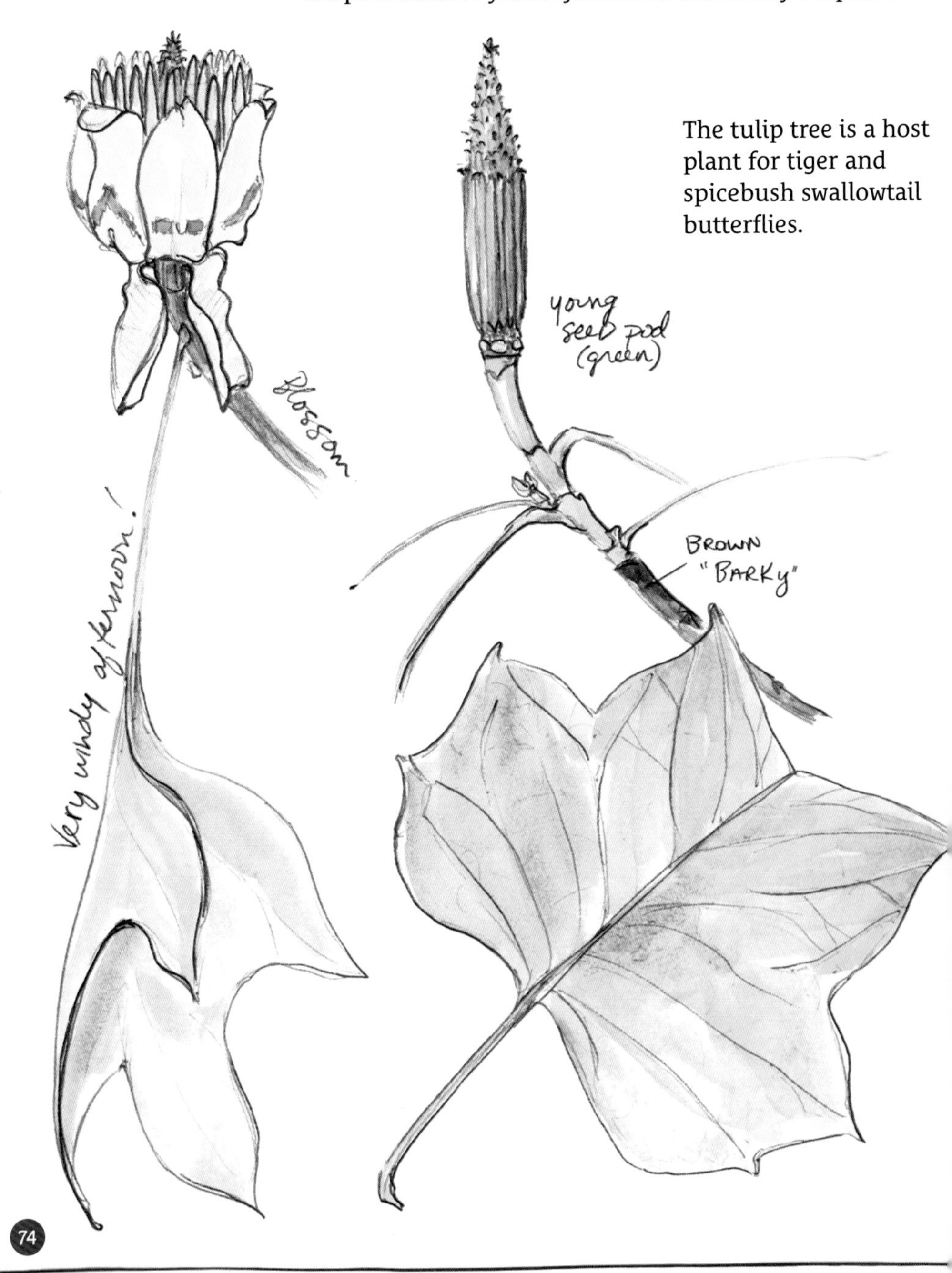

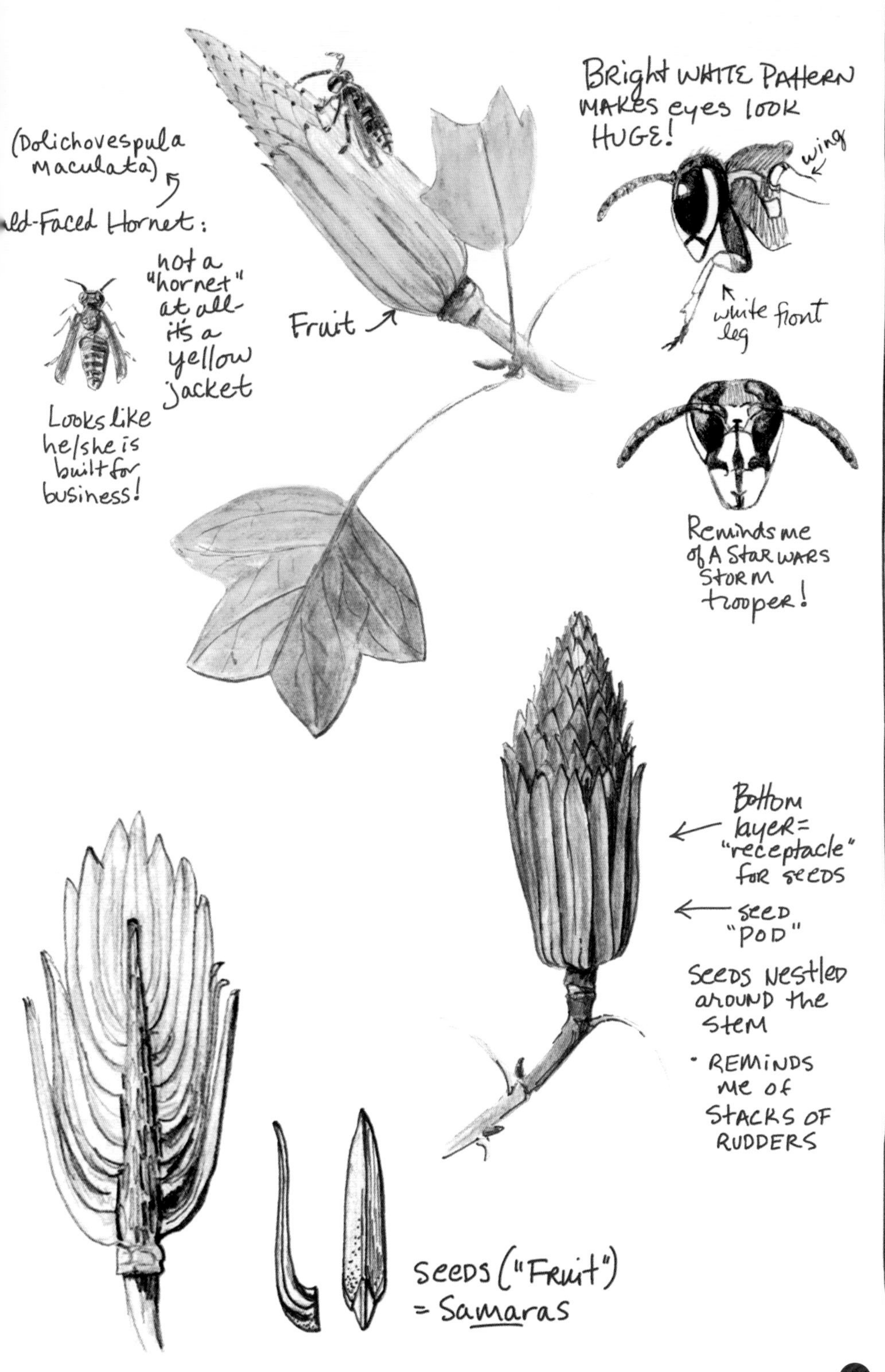
Bright WHITE PATTERN MAKES eyes look HUGE!
wing
(Dolichovespula Maculata)
ld-Faced Hornet:
not a "hornet" at all- it's a yellow jacket
Looks like he/she is built for business!
Fruit
white front leg
Reminds me of A STAR WARS STORM trooper!
Bottom layer = "receptacle" for seeds
seed "POD"
seeds Nestled around the stem
• REMINDS me of STACKS OF RUDDERS
seeds ("Fruit") = Samaras

Wych Elm

(*Ulmus glabra*)
June 28, 2021

At first glance, bundles of elm seeds look like young leaves or flowers. What a delight to discover pink seeds nestled in paper-thin bright green bracts.

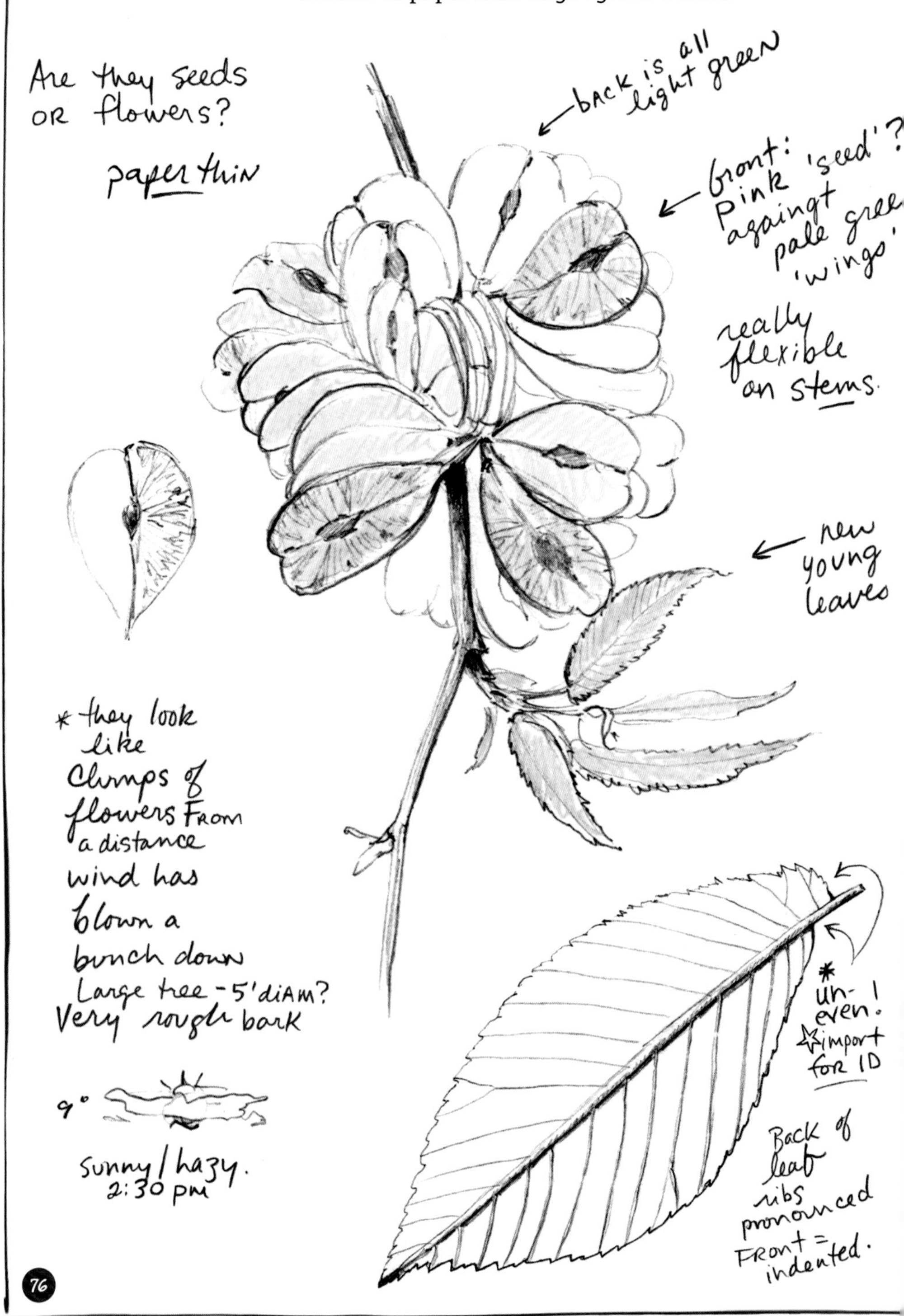

Western Red Cedar

(*Thuja plicata*)
May 7, 2016

The sun bouncing off this cedar reflected an amazing palette of colours in Ravine Park, four blocks west of Quilchena. As I wondered how to best draw the bark, a moth brushed my head and found a resting place on the tree in front of me. It was so well camouflaged! If I looked away for a second, it was hard to find again.

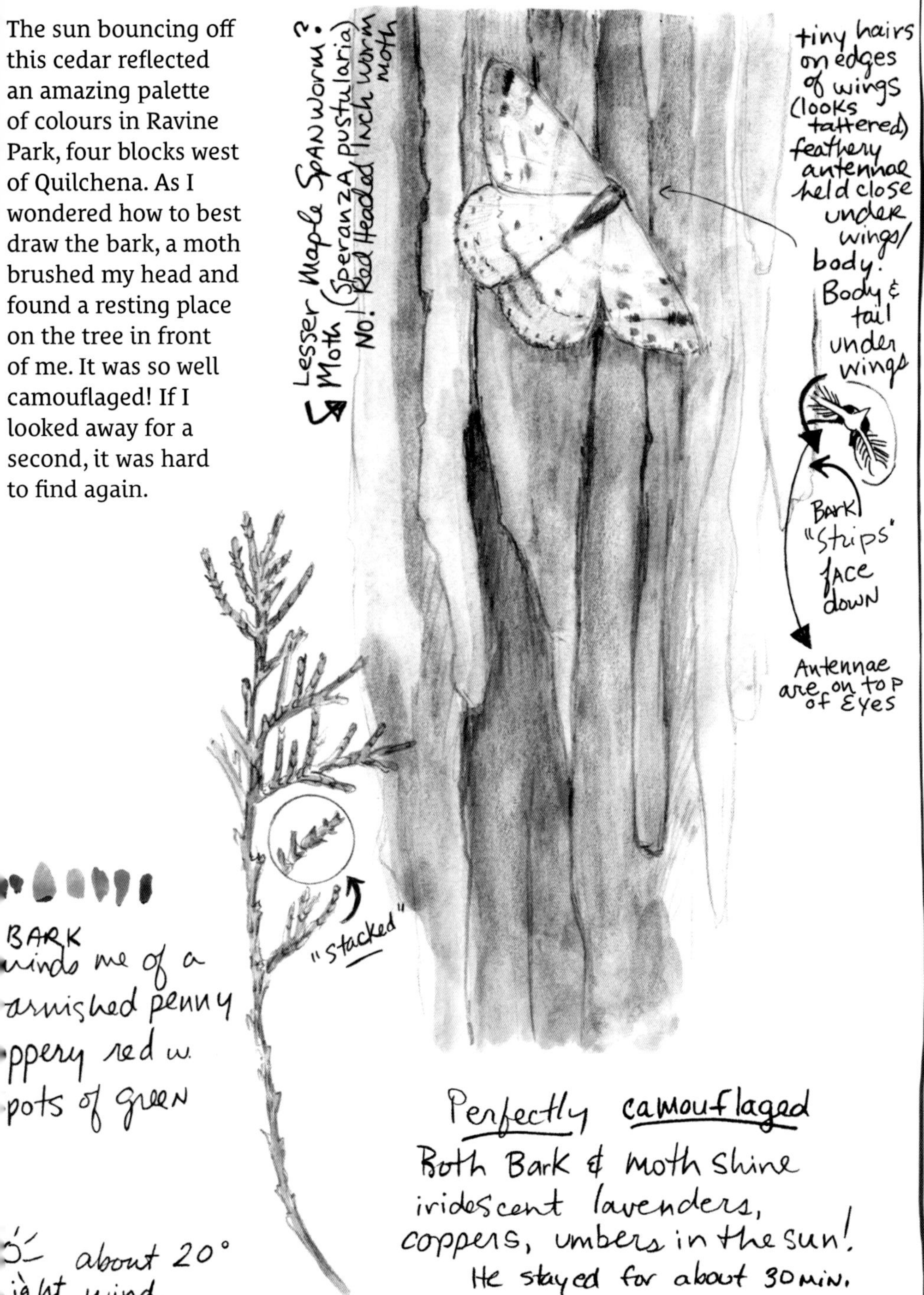

Swainson's Thrush

(*Catharus ustulatus*)
October 30, 2020

This unassuming-looking bird has a sweet, slightly sad song that spirals up in triplets. Swainson's thrushes like to forage in leaf litter on the forest floor – so it was a bit surprising to see this one on a branch in the city.

Witch's Butter

(*Tremella mesentericá*)

March 19, 2022

RAVINE PARK
MARCH 19, 2022
2:30pm 8C

Tremella mesenterica)
to "tremble" "mesentary" (intestines)
All sources say it's edible

- Usually found on dead deciduous trees or recently fallen branches.
- All *Tremella* species are parasites of crust fungus on rotting wood.
- The gelatinous "jelly" blobs are the fruiting bodies of the fungus.
- Other common names include: golden jelly fungus and yellow trembler.

Skunk Cabbage

(*Lysichiton americanus*)
March 21, 2020

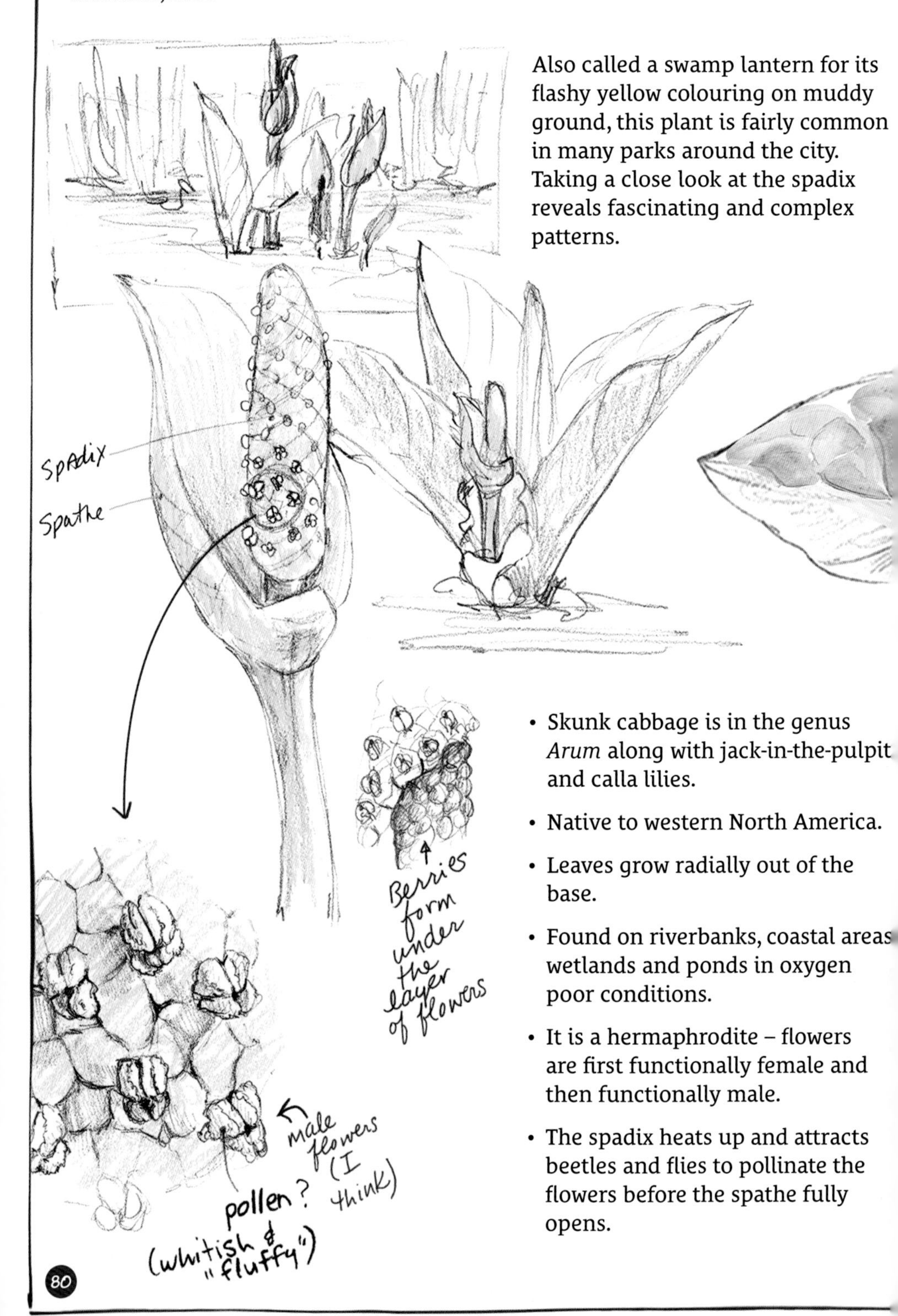

Also called a swamp lantern for its flashy yellow colouring on muddy ground, this plant is fairly common in many parks around the city. Taking a close look at the spadix reveals fascinating and complex patterns.

- Skunk cabbage is in the genus *Arum* along with jack-in-the-pulpit and calla lilies.
- Native to western North America.
- Leaves grow radially out of the base.
- Found on riverbanks, coastal areas wetlands and ponds in oxygen poor conditions.
- It is a hermaphrodite – flowers are first functionally female and then functionally male.
- The spadix heats up and attracts beetles and flies to pollinate the flowers before the spathe fully opens.

SAKURA
Lemon Yellow + white.
- Blue violet
+ Lemon
CANARY PRISMA.

3pm. Clear
11°C. Ravine
PARK
Vancouver

Volunteer and Tatlow Parks

Tatlow, along with Volunteer Park located directly across Point Grey Road to the north, are small areas of quiet green space tucked into the Kitsilano neighbourhood. Overlooking English Bay and the North Shore Mountains, these parks are the site of a historical stream, originally called First Creek, that was buried during the industrialization of Vancouver. A small section of the stream is still visible running through Tatlow Park before it is diverted to underground pipes. In 2017, plans got under way to restore this stream in a process called "daylighting." This will reconnect the stream above ground in Tatlow, through Volunteer Park to English Bay. While the stream is not expected to become active for salmon spawning, the goal is to increase biodiversity by adding native plants and restoring stream banks to create a thriving riparian habitat.

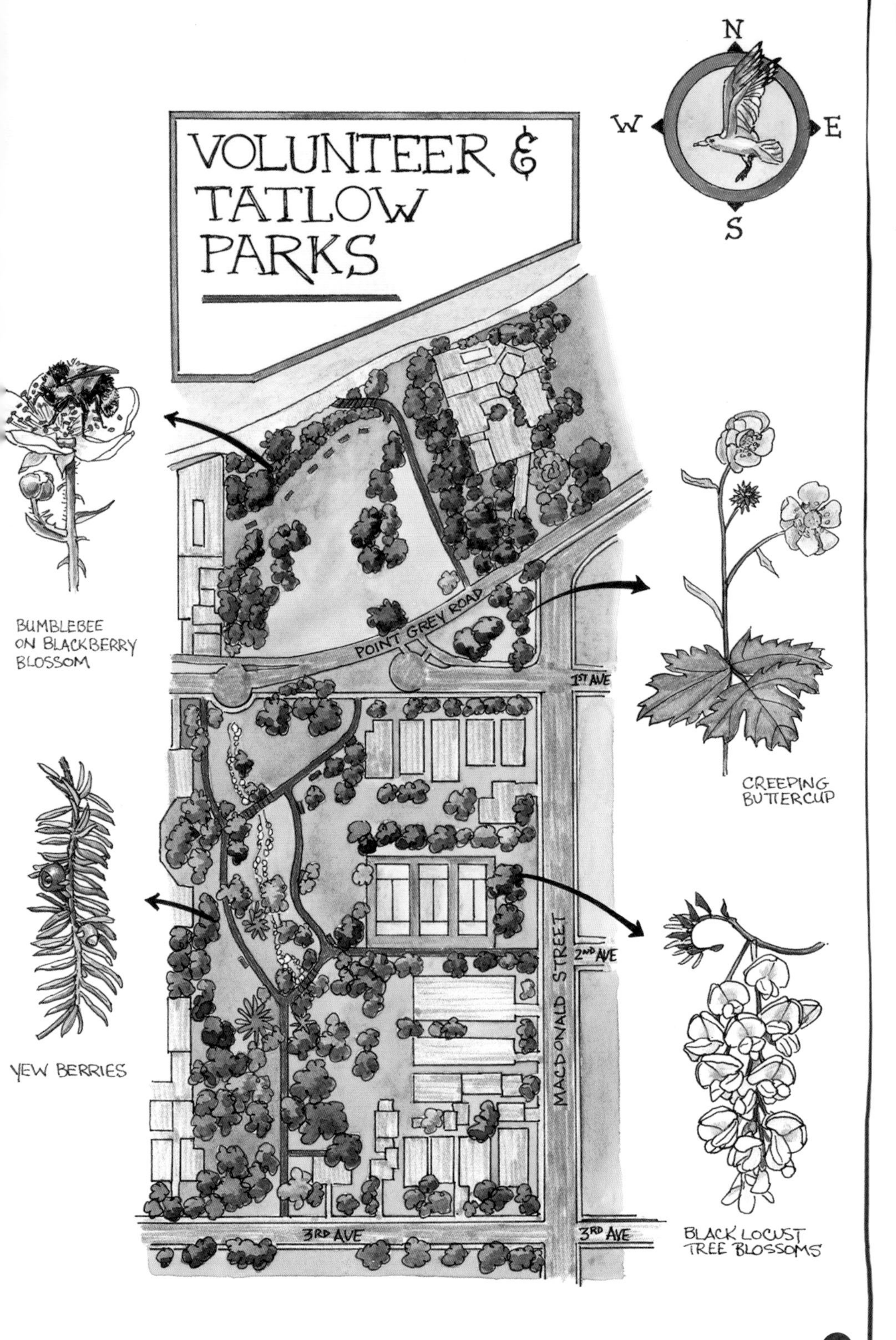

VOLUNTEER & TATLOW PARKS
N
W
E
S
BUMBLEBEE ON BLACKBERRY BLOSSOM
CREEPING BUTTERCUP
YEW BERRIES
BLACK LOCUST TREE BLOSSOMS
POINT GREY ROAD
1ST AVE
2ND AVE
3RD AVE
3RD AVE
MACDONALD STREET

Ocean Gems at Your Feet

October 27, 2021

VOLUNTEER PARK

• Strolling down the stairs at the East end of the park is a small stretch of beach.....

...with countless treasures Did not know we have true oysters here in the ci

interesting stone

SHIELD LIMPET

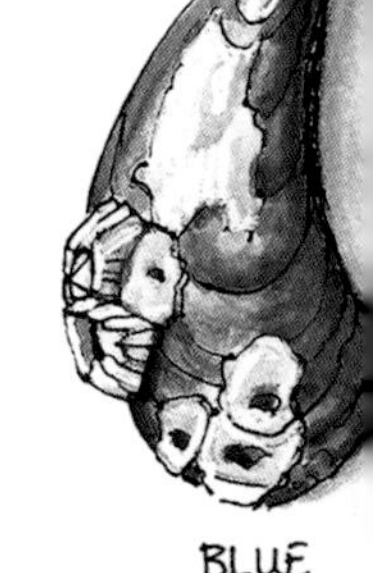

BLUE MUSSEL WITH BARNAC

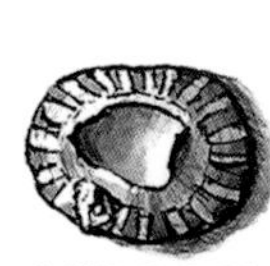

CHECKERED LIMPET

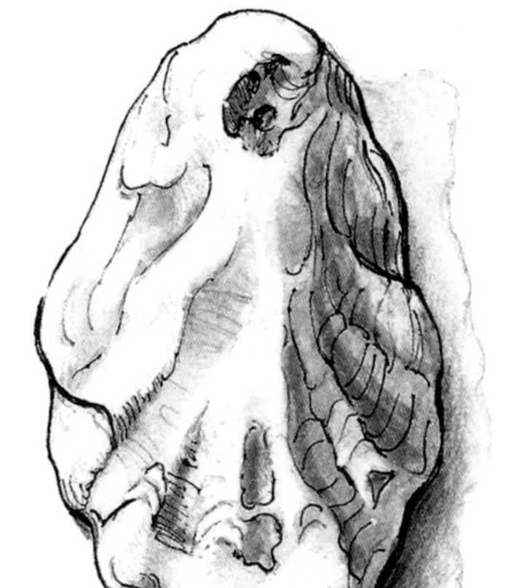

JAPANESE (PACIFIC) OYSTER

MANILA CLAM

ooking North East towards
e city – this is as far as
ou can go at high tide.

- Venus / MANILA CLAMS
- JAPANESE / PACIFIC OYSTERS
- BLUE MUSSELS
- SHIELD LIMPETS.

VENUS
CLAM

TINY
DUNGENESS CRAB
(MOULT)

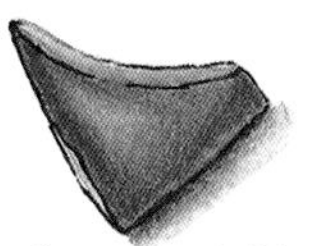

SEA GLASS

PERIWINKLE

BLUE MUSSEL

PURPLE MAHOGANY CLAM

White Cabbage Butterfly

(*Pieris rapae*)
August 25, 2020

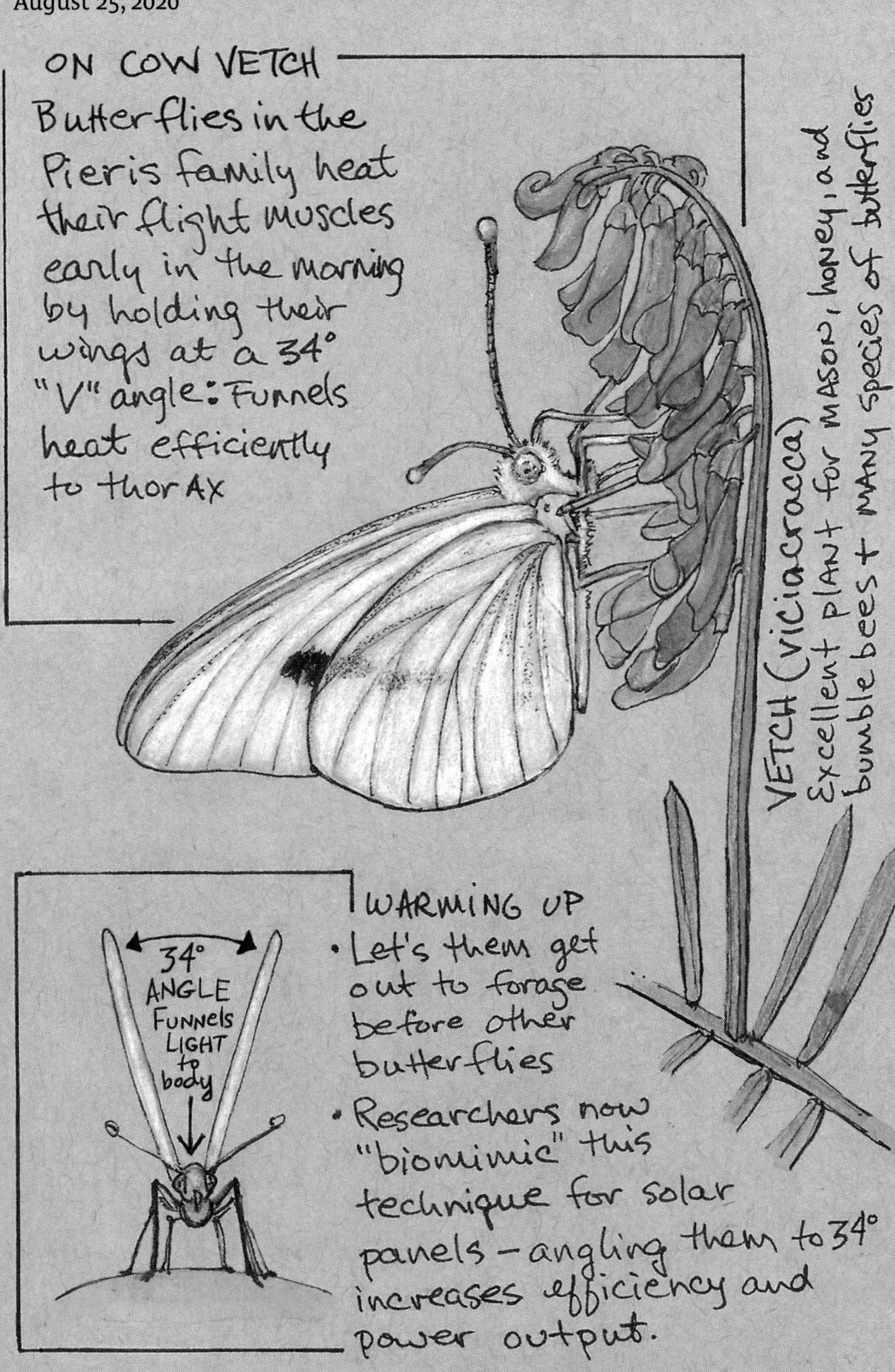

Rufous Hummingbird

(*Selasphorus rufus*)
May 13, 2021

It's always a delight to see a tiny rufous hummingbird! They migrate up to 4,830 kilometres (3,000 miles) from wintering grounds in Mexico to British Columbia and Alaska. The sound of their buzzing wings always reminds me of miniature helicopters. Brilliant oranges and reds flashing off a male's gorget – the patch of colour on the bird's throat – are both a visual treat and a challenge to paint.

Bewick's Wren

(*Thryomanes bewickii*)

December 13, 2021

- Light coloured "eyebrow" is good I.D.
- Usually shy & speedy.
- Common all year around Vancouver
- Males learn their songs before their 1st winter & keep it their entire lives.
- Look for them in understory

☆ A group of wrens is called a "chime", a "flight" and a "flock."

6°C

Apple of Peru

(*Nicandra physalodes*)
October 27, 2021

- Also Known as "Shoo-fly" plants
- Native to S. America
- Member of Solanaceae = nightshade/potato family

2:30pm
overcast 10°C

- So intriguing & omenous at the same time.
- A perfect find for Halloween

- REMINDS ME OF BAT WINGS!

Stanley Park

Vancouver's oldest and most spectacular urban park, Stanley Park takes up the entire peninsula of land immediately west of downtown Vancouver. It is home to a great diversity of plants, birds, and animals, with natural habitats that include conifer forests, boggy wetlands, and rocky shores. Surrounded by Burrard Inlet, Coal Harbour, and English Bay, the park is situated on traditional territory of the Coast Salish First Nations, including Musqueam, Squamish, and Tsleil-Waututh unceded lands. It was home to the Squamish villages Whoi Whoi and Chaythoos on the park's northern shore; to Hawaiian immigrants who lived at Kanaka Ranch near the park's entrance; and to a Squamish-Portuguese community at Brockton Oval.

Established as Vancouver's first city park in 1888, it remains one of the largest urban parks in Canada today. Garden designs and features were blended with natural elements by the City of Vancouver between 1913 and 1936. W.S. Rawlings, the Vancouver Board of Parks and Recreation's first superintendent, designed gardens, landscapes, and recreational facilities to enhance the visitor experience. The Vancouver Aquarium, miniature train, and children's zoo were added during the post-war period from 1950 to 1973.

Stanley Park continues to be a cultural site, commemorating notable people and events such as writer E. Pauline Johnson; Olympic athlete Harry Jerome; adventurer Joe Silvey, better known as "Portuguese Joe"; the Vancouver Centennial; Japanese Canadian soldiers in WWI; the Salvation Army; and shipwrecks of the *Chehalis, Beaver,* and HMS *Egeria,* among others.

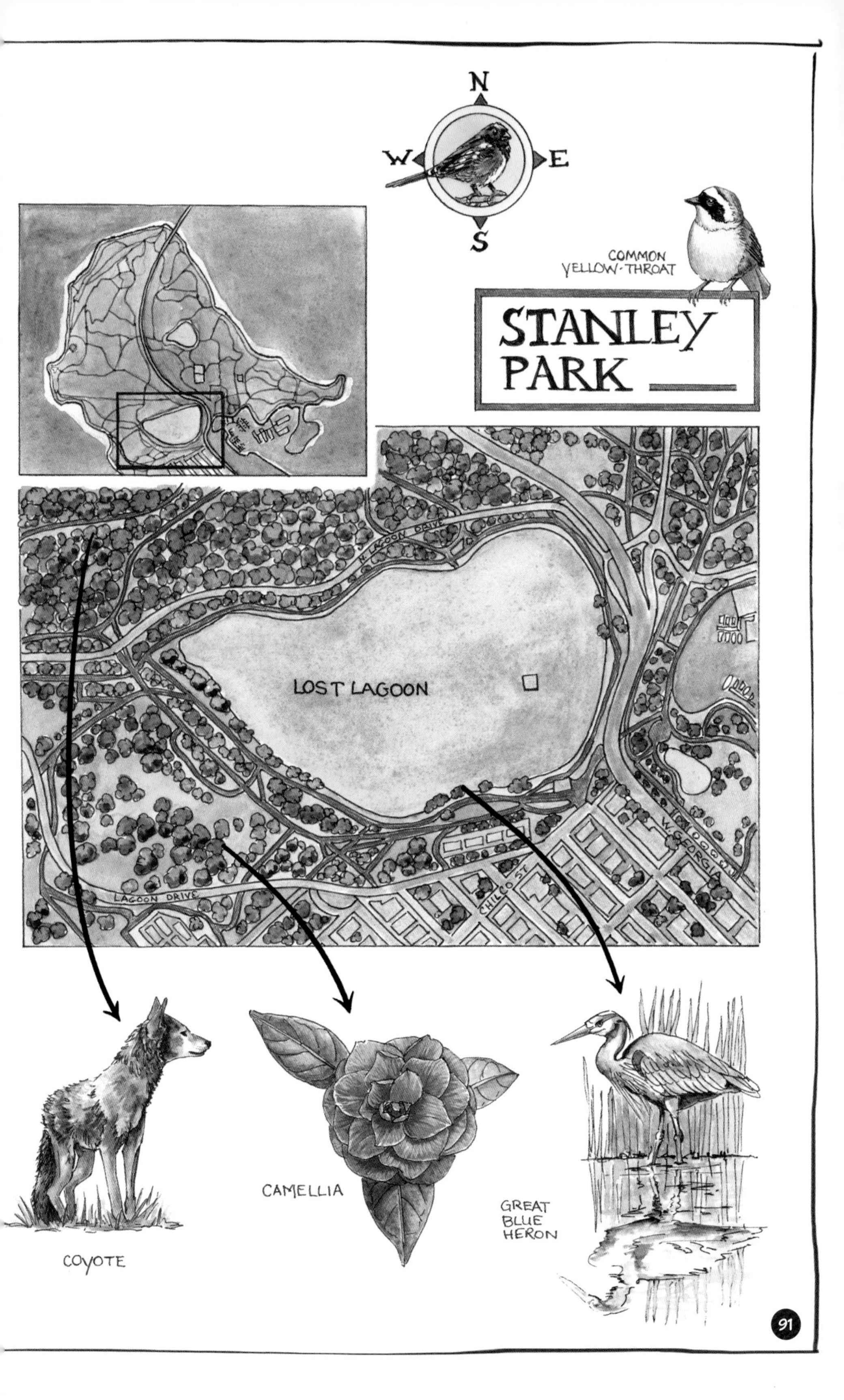
N
W
E
S
COMMON
YELLOW-THROAT
STANLEY
PARK
N LAGOON DRIVE
LOST LAGOON
LAGOON DRIVE
CHILCO ST
W. GEORGIA
COYOTE
CAMELLIA
GREAT
BLUE
HERON

Raccoon

(*Procyon lotor*)
April 9, 2022

Did you know...

- Raccoons are experts at quickly adapting to changing environments. That's one reason why they have thrived for forty thousand years, especially in urban settings.
- A raccoon can squeeze its whole body through a hole only 4.6 centimetres (3 inches) across. (That's the size of a baseball.)
- Raccoons can descend a tree headfirst by rotating their hind feet 180 degrees.
- Many people believe that raccoons wash their food before they eat it, but this is a myth. Raccoons have ten times as many nerve endings in their hands as we do, and four to five times as many mechanoreceptors than other animals. Mechanoreceptors are sense organs that respond to stimuli like touch and vibration. Water increases the sensitivity of these mechanoreceptors, which allows raccoons to identify edible food items more accurately.
- Raccoons can damage homes and gardens, but by understanding their seasonal activities we can minimize or resolve human-animal conflict before it happens.
- Raccoons typically sleep during the day and are most active at night. Surprisingly, I spotted this young raccoon mid-day by Lost Lagoon. It waded through the shallows at the edge of the water, crossed a log, and clambered up a tree right next to the pedestrian pathway. I did a few quick gesture drawings of him in the park, but completed this "peek-a-boo" sketch back in the studio.

Ruby-Crowned Kinglet

(*Corthylio calendula*)

April 9, 2022

Little Brown Bat

(*Myotis lucifugus*)

April 9, 2022

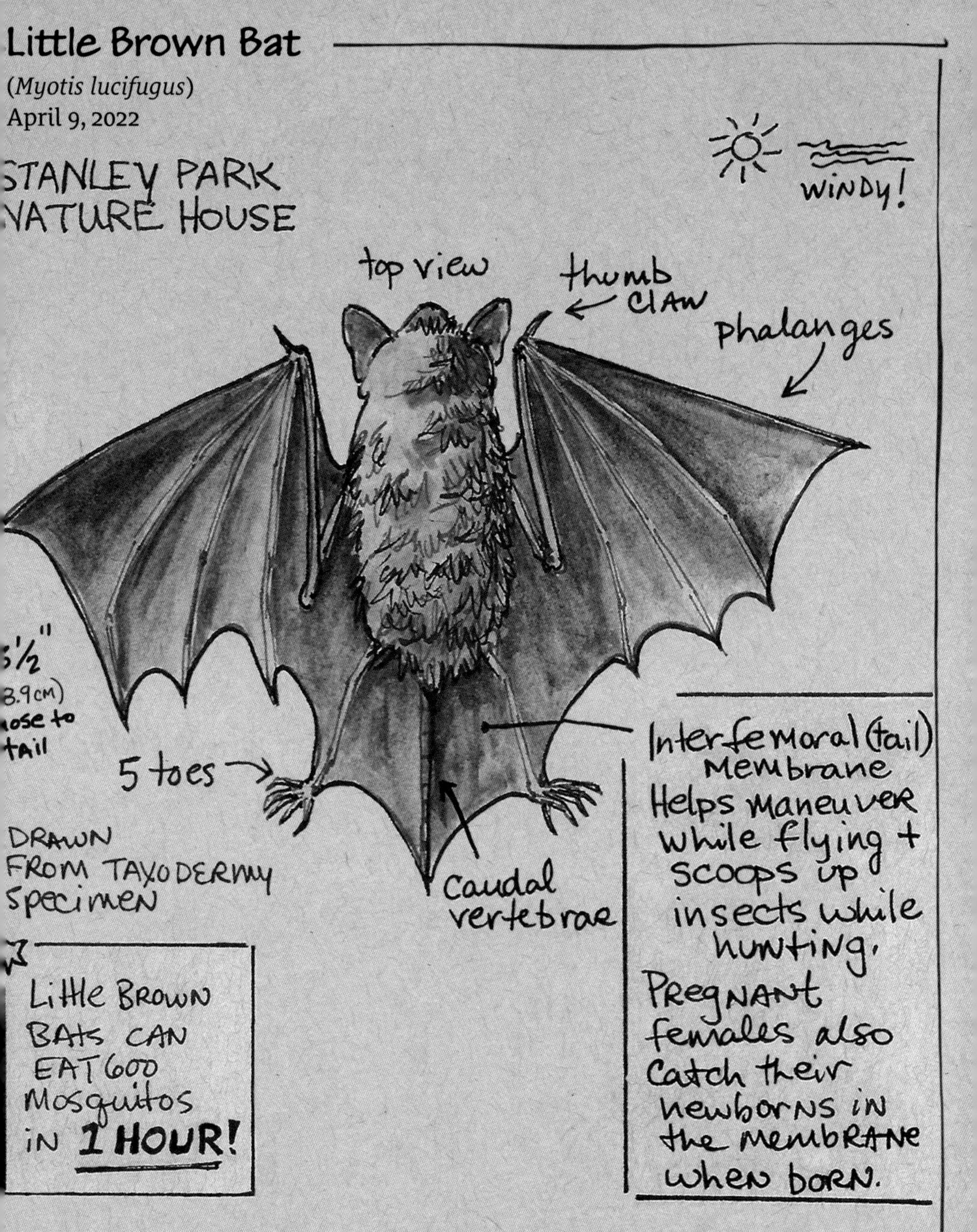

The little brown bat was emergency listed as endangered on the federal Species at Risk Act in 2014 because of sudden population declines. Half of the sixteen bat species in British Columbia are listed as vulnerable or threatened. Bats are vital for the health of our environment – important for controlling pests, pollinating plants, dispersing seeds, and keeping our ecosystem in balance. One little brown bat can eat up to six hundred mosquitoes in just one hour!

Six species of bats live in Stanley Park: Yuma myotis, silver-haired bat, California myotis, little brown bat, hoary bat, and big brown bat. Look for them when they emerge to hunt insects in the early evening at dusk.

Wood Duck

(*Aix sponsa*)
April 7, 2022

BREEDING DISPLAY?
OR JOYFUL STRETCHING?

Wilson's Warbler

(*Cardellina pusilla*)
May 18, 2021

I first noticed bright-yellow flashes darting in and out of the Nootka rose bushes. Such a shy bird, but an absolute delight to see, Wilson's warblers sport some of the most vivid yellow of all birds in this group. They are also one of the smallest, weighing the equivalent of two toonie coins. A group of warblers are called a "confusion," a "wrench," or a "bouquet."

Black-Capped Chickadees

(*Poecile atricapillus*)

February 17, 2020

6°C

atricapillus)
= black (latin) = head (latin) = hair (latin)

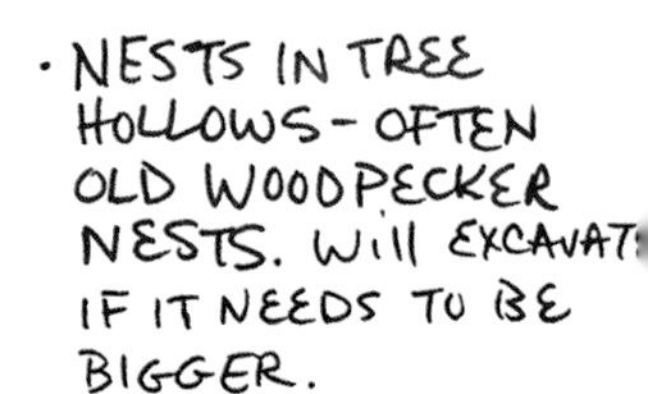

• NESTS IN TREE HOLLOWS – OFTEN OLD WOODPECKER NESTS. WILL EXCAVATE IF IT NEEDS TO BE BIGGER.

• CHICKADEES USUALLY MATE FOR LIFE

• THEY HAVE A VERY COMPLEX LANGUAGE – HAVING RULES FOR HOW VARIOUS NOTES ARE ORDERED

• THE "D" NOTE IN "CHICK·A·DEE·DEE·DEE" CALL IS USED WHEN A NEW FOOD SOURCE IS FOUND, MOVEMENT IS NEEDED OR PREDATORS ARE NEAR. THE MORE "DEES" THE MORE DANGEROUS THE PREDATOR OR THE NEWER THE FOOD SOURCE. BOTH INSTANCES ALERT THE FLOCK TO BAND TOGETHER.

Dee Dee Dee

• OTHER CHICKADEE CALLS INCLUDE GURGLES, THE "FEE BEE" SONG (usually FOR TERRITORY/FINDING A MATE), SOFT SINGLE TWEETS & CHATTERS.

Red-Eared Slider

(*Trachemys scripta elegans*)
April 9, 2022

°C 4:45pm

- subspecies of the pond slider.

- A patch of red behind each eye gives these turtles their common name.
- A thin membrane covers their ears – they can actually hear better underwater. Water serves as a conductor and amplifies low-frequency sound vibrations.
- Red-eared sliders are native to the Mississippi River in the United States. They are a popular pet shop species in Canada that were likely released into local ponds and wetlands.
- These turtles have overwhelmed ecosystems throughout British Columbia and Ontario and have displaced the only native freshwater turtle in our province – the Western painted turtle – which is now listed as an endangered species.

Queen Elizabeth Park

Once known as "Little Mountain," Queen Elizabeth Park is located at the geographic centre of Vancouver and is the city's highest point, reaching just over 152.5 metres (501 feet) above sea level. Originally forested with old-growth conifers, the area's basalt outcroppings were of interest to the Canadian Pacific Railway, and they took possession in the late 1800s. Between 1890 and 1911, basalt rock mined within the park was used to build the first roadways in Vancouver. Canadian Pacific Railway reserved the land specifically for park purposes beginning in 1912. After Vancouver amalgamated with the municipalities of Point Grey and South Vancouver in 1929, the City of Vancouver officially acquired this property.

In July 1940, Little Mountain was renamed Queen Elizabeth Park in dedication to a visit from the royal couple the previous year. With $5,000 per year funding from the Canadian Pulp and Paper Association, Park Board staff began transforming the overgrown hillsides of the park into Canada's first urban arboretum. Examples of native trees found across Canada, including Douglas and amabilis firs, cedars, and black cottonwood, were planted alongside many exotic species to create the park we know today.

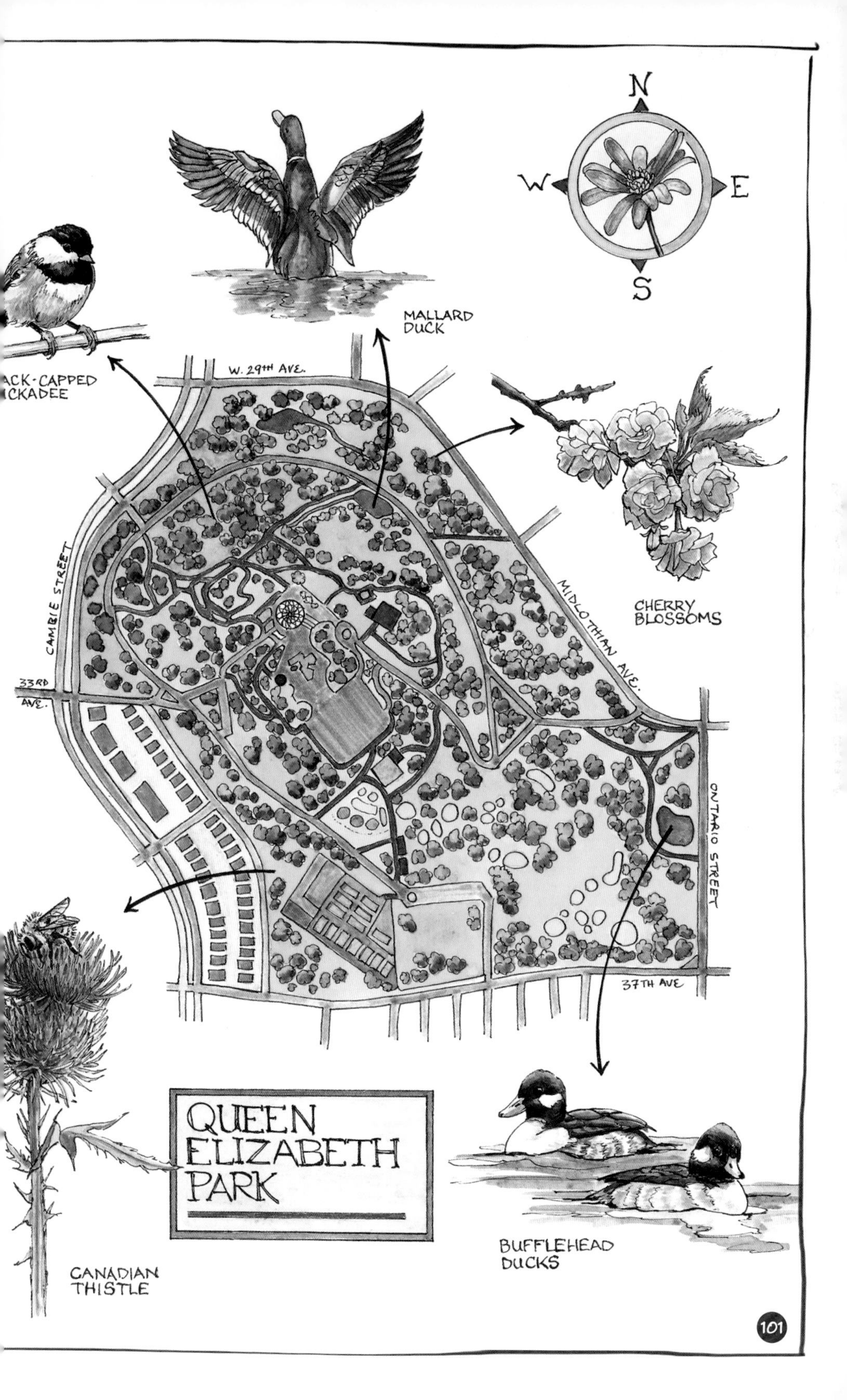
N
W
E
S
MALLARD DUCK
ACK-CAPPED
CKADEE
W. 29TH AVE.
CHERRY BLOSSOMS
CAMBIE STREET
MIDLOTHIAN AVE.
33RD AVE.
ONTARIO STREET
37TH AVE
QUEEN ELIZABETH PARK
BUFFLEHEAD DUCKS
CANADIAN THISTLE

Brown Creeper

(*Certhia americana*)
November 8, 2020

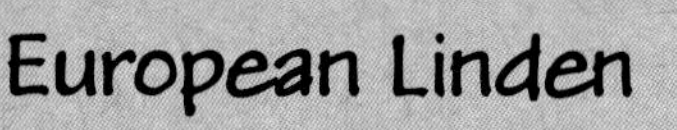

(*Tilia x europaea*)
November 8, 2020

Flowering Dogwood

(*Cornus florida*)
July 21, 2021

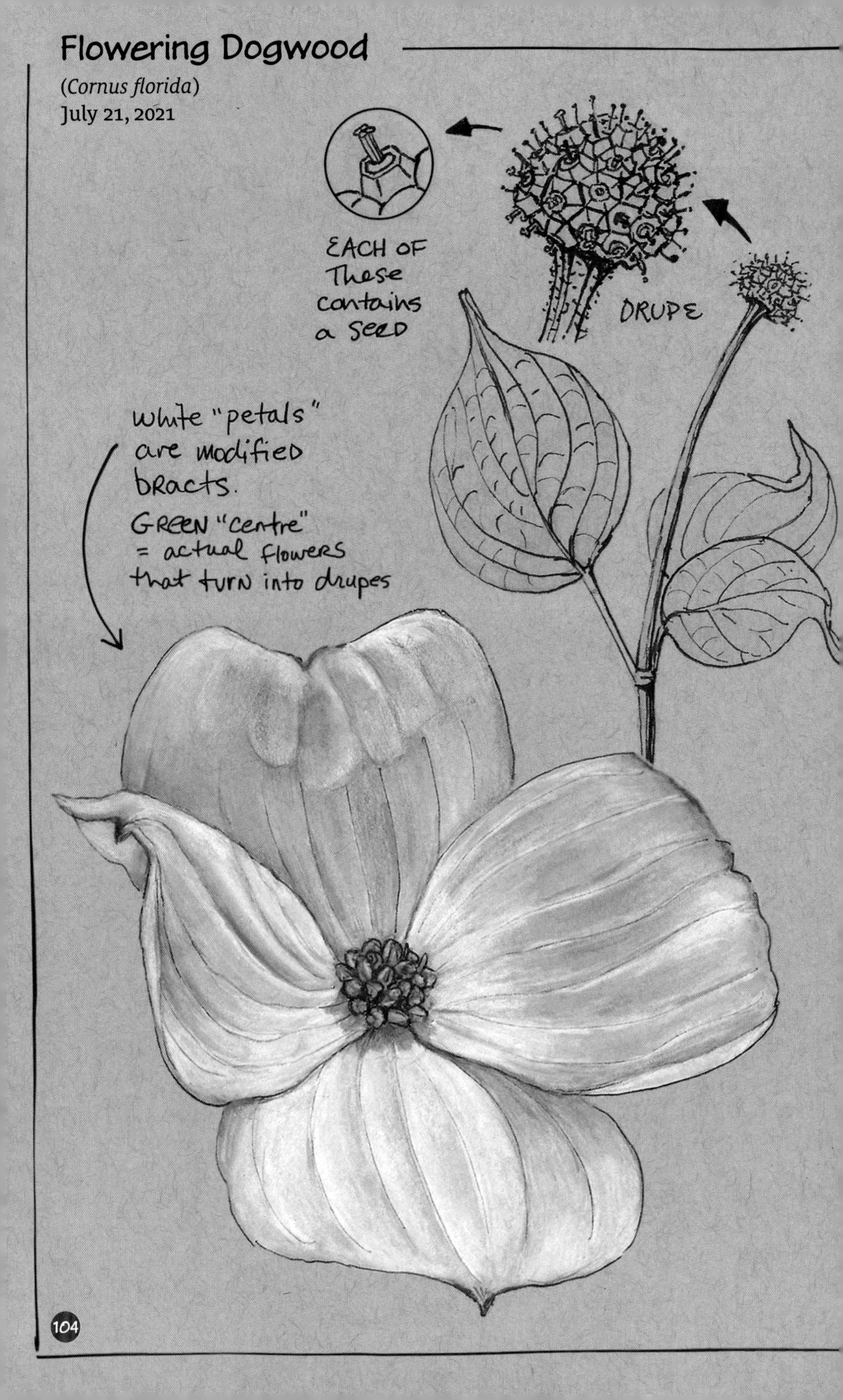

Oak Lantern Moth

(*Carcina quercana*)
July 21, 2021

This tiny jewel-coloured moth surprised me as I tried to get a better look at chickadees rummaging in the foliage. Before it disappeared under a leaf, I noticed that the moth's antennae were held forward like pincers. Oak lantern moths are an invasive species. Also called the "oak skeletonizer," they can cause severe damage to oak trees by stripping their leaves and reducing growth.

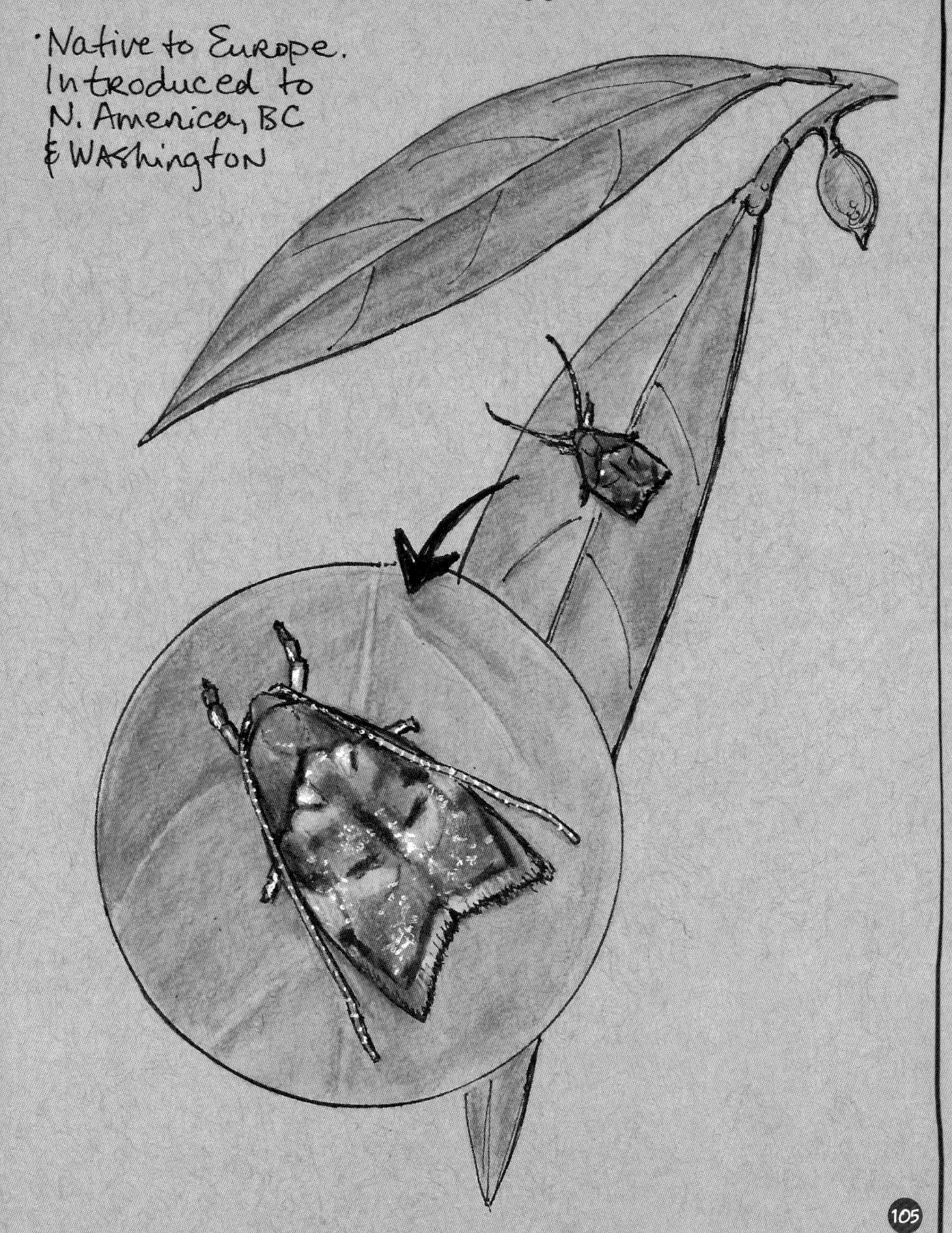

Spotted Towhee

(*Pipilo maculatus*)
February 17, 2020

One of my favourite birds, towhees can usually be found foraging in the underbrush looking for insects. Their bright-red eyes give them a slightly disgruntled or stern look if they spot you watching them.

Golden-Crowned Kinglet

(Regulus satrapa)
November 3, 2019

Golden-crowned Kinglet
Flock moving through
QE PARK (N.) &
eeDing on berries
" (7.6 cm) in length.
ommon year round
n Vancouver & up/
down entire west
oast of N. America

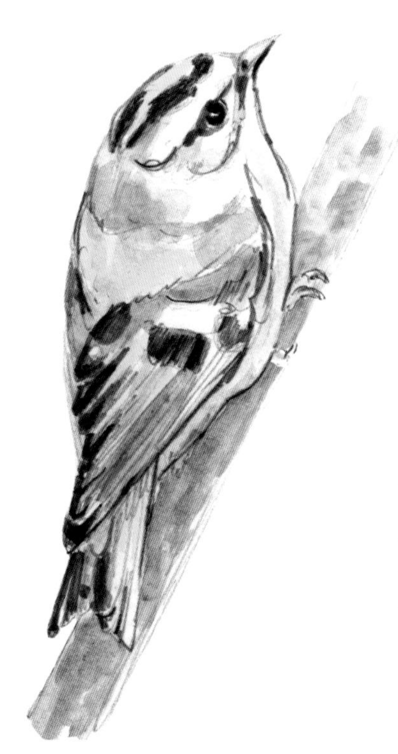

Kinglets are tiny, quick birds that eat mainly insects. The yellow crest of the golden-crowned kinglet can be easily seen compared to the red crest of the ruby-crowned kinglet which is usually hidden. Golden-crowned kinglets are found mainly around conifer trees and can join in with flocks of warblers.

Common Hazelnut

(*Corylus avellana*)
August 15, 2020

The elegant, flowing husks around the seeds of this hazelnut caught my eye and reminded me of both fire and water. Also called filberts, hazelnuts are high in nutrients and prized by squirrels, birds, and humans alike. When the nuts drop from their trees in the city, they disappear almost as soon as they hit the ground.

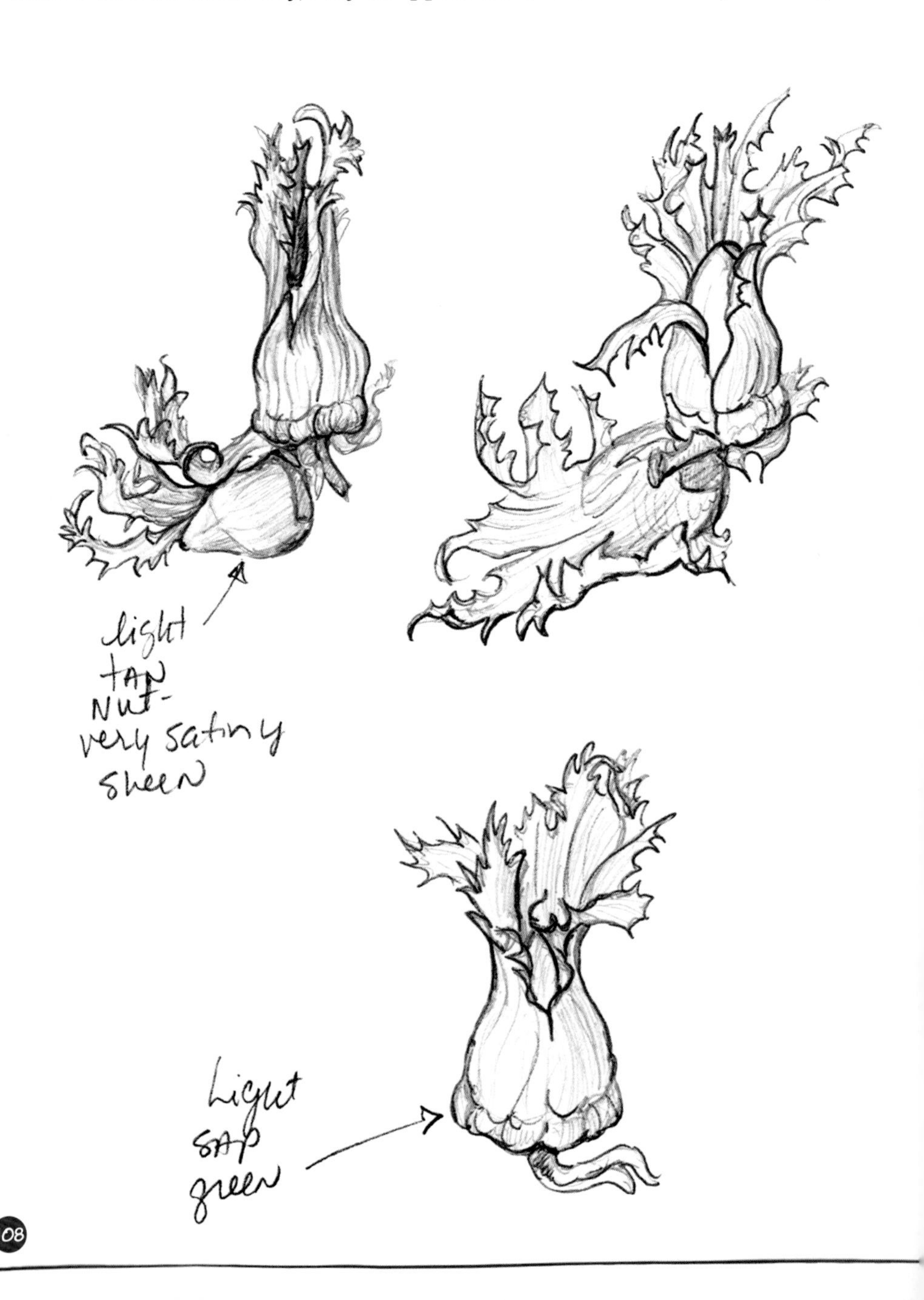

European Larch

(*Larix decidua*)
March 16, 2021

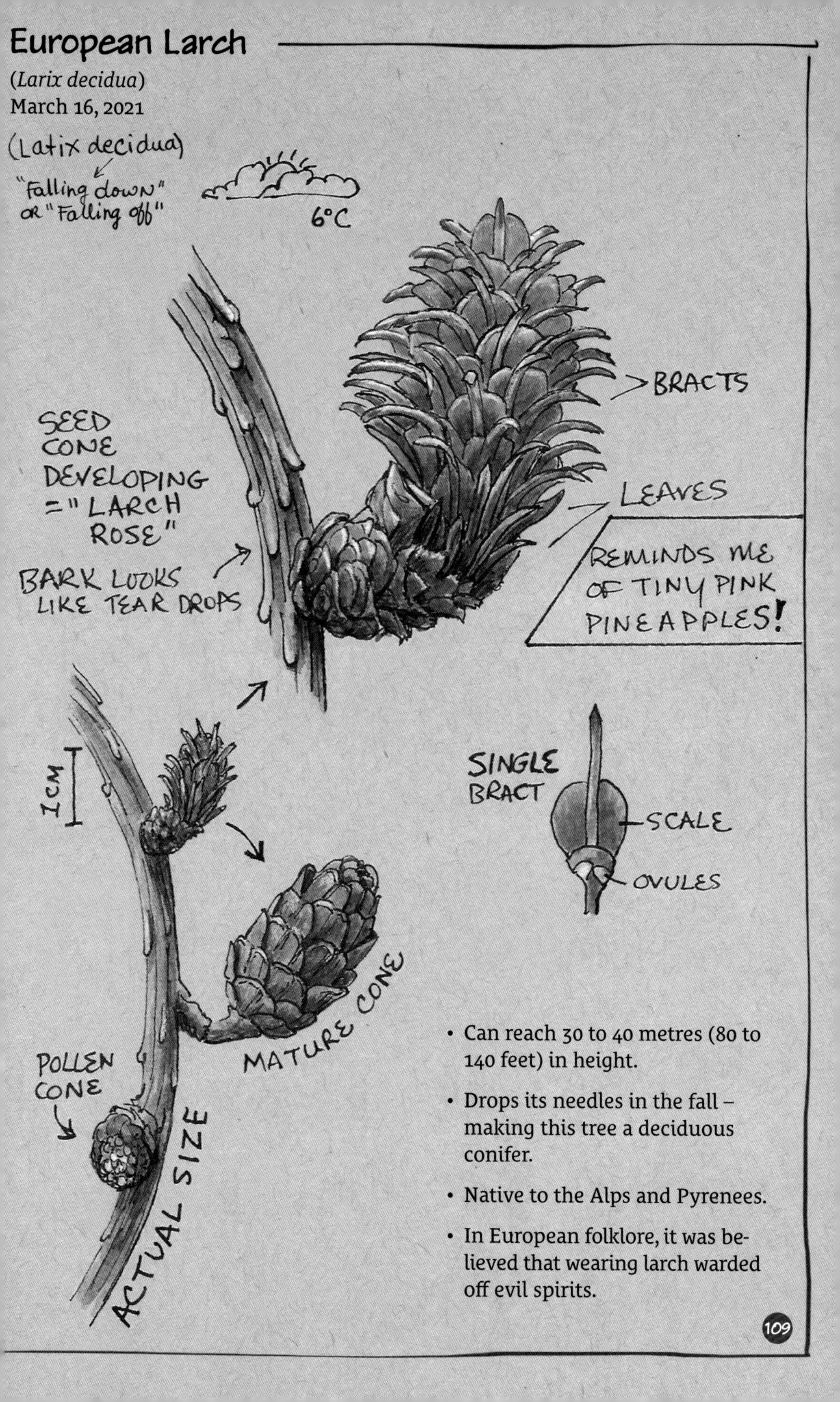

- Can reach 30 to 40 metres (80 to 140 feet) in height.
- Drops its needles in the fall – making this tree a deciduous conifer.
- Native to the Alps and Pyrenees.
- In European folklore, it was believed that wearing larch warded off evil spirits.

Trout Lake (John Hendry Park)

Trout Lake was originally a peat bog lake that supported salmon, trout, and beaver. In the 1860s, the lake supplied water to the boilers at Hastings Mill lumber company on Burrard Inlet until the Capilano water system was developed in 1880. Vancouver's Board of Parks and Recreation began acquiring adjacent land in 1920. In 1926, Aldyen Hendry Hamber – daughter of John Hendry, the owner of Hastings Mill – donated a portion of the land at Trout Lake to the City of Vancouver under the condition it be renamed John Hendry Park. By 1932, the Park Board began mining the peat in the lake for revenue. A field house was built in 1951, after which the lake became a popular place to swim in the summer and skate in the winter. Other amenities were added over the years, including recreation facilities, art studios, a performance plaza, and a popular summertime farmer's market. As one of Vancouver's few freshwater lakes with adjacent riparian habitat, it is frequented by a rich number of bird species making Trout Lake a "hotspot" for bird watchers.

TROUT LAKE
AKA JOHN HENDRY
PARK
N
E
W
S
TEMPLETON DR.
14TH
AVE
E. 19TH AVE.
BEWICK'S
WREN
FLICKER
STELLER'S JAY
BUSHTIT
NORTHERN SHOVELER

A Day at the Lake

March 5, 2022

BEWICK'S WR
BUSHTITS
GOLDEN-CROWNED KINGLET
GOLDFINCHE
CHICKADEES
FLICKER
CROWS
CANADA GEESE
ROBINS
STELLER'S JA
AMERICAN WIGEONS
WOODDUCKS
N. SHOVELERS
BALD EAGLE
COOTS
SEAGULLS
MALLARDS
↑
LOTS OF BIRDS ON/AROUND LAKE!

AMERICAN WIGEON - FEMALE & MALE
(*Mareca americana*)

American Goldfinch

(*Spinus tristis*)

- Heavily dependent on thistle, sunflowers, asters, seeds of alder, birch, and western cedar for food.
- Often uses thistledown to line their cup-shaped nests.
- Attaches nests to supporting twigs with spider web silk.

Steller's Jay

(*Cyanocitta stelleri*)
October 19, 2019

(Cyanocitta stelleri)
"blue" "jay" "tall black crest"

- RANGE IS FROM ALASKA TO NICARAGUA
- DOESN'T MIGRATE.
- LOUD & NOISY - THEY CAN IMITATE CALLS OF OTHER BIRDS & ANIMALS

- JAY WITH ACORN
- THEY OFTEN CACHE SEEDS/NUTS IN THE GROUND & IN TREES TO EAT LATER
- HIGHLY SOCIAL BIRDS.

Canada Goose

(*Branta canadensis*)
May 17, 2022

The Canada goose is a familiar sight in Vancouver – so common most don't give them a second thought. They are a majestic bird that is easily identified by their large size and their white cheek patch, called a chin strap. They bond for life and are devoted partners and parents. While Canada geese usually migrate south for the winter, our city's climate is suitable for them to stay all year-round.

I am always drawn to the grace and elegance of a bird's wings and feathers. This goose was having a stretch after preening – a time-consuming task! Preening is vital for birds. Feathers must be clean and dry so that they can fly at a moment's notice. Each individual feather is tended to one by one by pulling it through the ridges on their bill to remove dirt and water. Shaking, ruffling, and flapping also gets rid of excess water and allows air to flow into the feathers.

American Coot

(*Fulica americana*)
March 20, 2022

I never paid much attention to this bird until I saw their feet! Now they are one of my favourite aquatic birds. Lime-green and turquoise-blue flexible, lobed feet have evolved to provide efficient travel over different surfaces. The open grid patterns remind me of fishnet stockings.

- Most Aquatic member of the RAil family
- CRAZY Cool lobed feet have evolved for efficient swimming & walking on mud, ice, grass.
- Lobes on feet fold back when they take a step on land & on the forward stroke in the water.

Everett Crowley Park

This medium-sized park located in the Champlain Heights area of South Vancouver was originally home to a mature Western hemlock and Western red cedar forest. A ravine cut north to south through the centre of the area and once contained a salmon-bearing stream that connected to the Fraser River. The area supported a diverse variety of birds, amphibians, reptiles, and larger mammals, including bear, elk, and mule deer.

In the 1870s, settlers began to log the area to use it for farming. In the 1920s, when forest harvesting greatly increased around the city, the Old Dominion and Canadian White Pine Mills opened near the south end of Boundary Road on the Fraser River. During the 1930s and 1940s, a sand and gravel quarry began operating on the north side of the park. Sixteen years later, the area became the city's waste deposit site. Once the quarry was abandoned in the 1970s, ground and surface water collected in what is now Avalon Pond. Meanwhile, the area began to recover as natural vegetation re-established itself.

The Everett Crowley Park Committee, made up of residents and park users, now works closely with the Vancouver Board of Parks and Recreation to support the park's ongoing natural recovery and reforestation process.

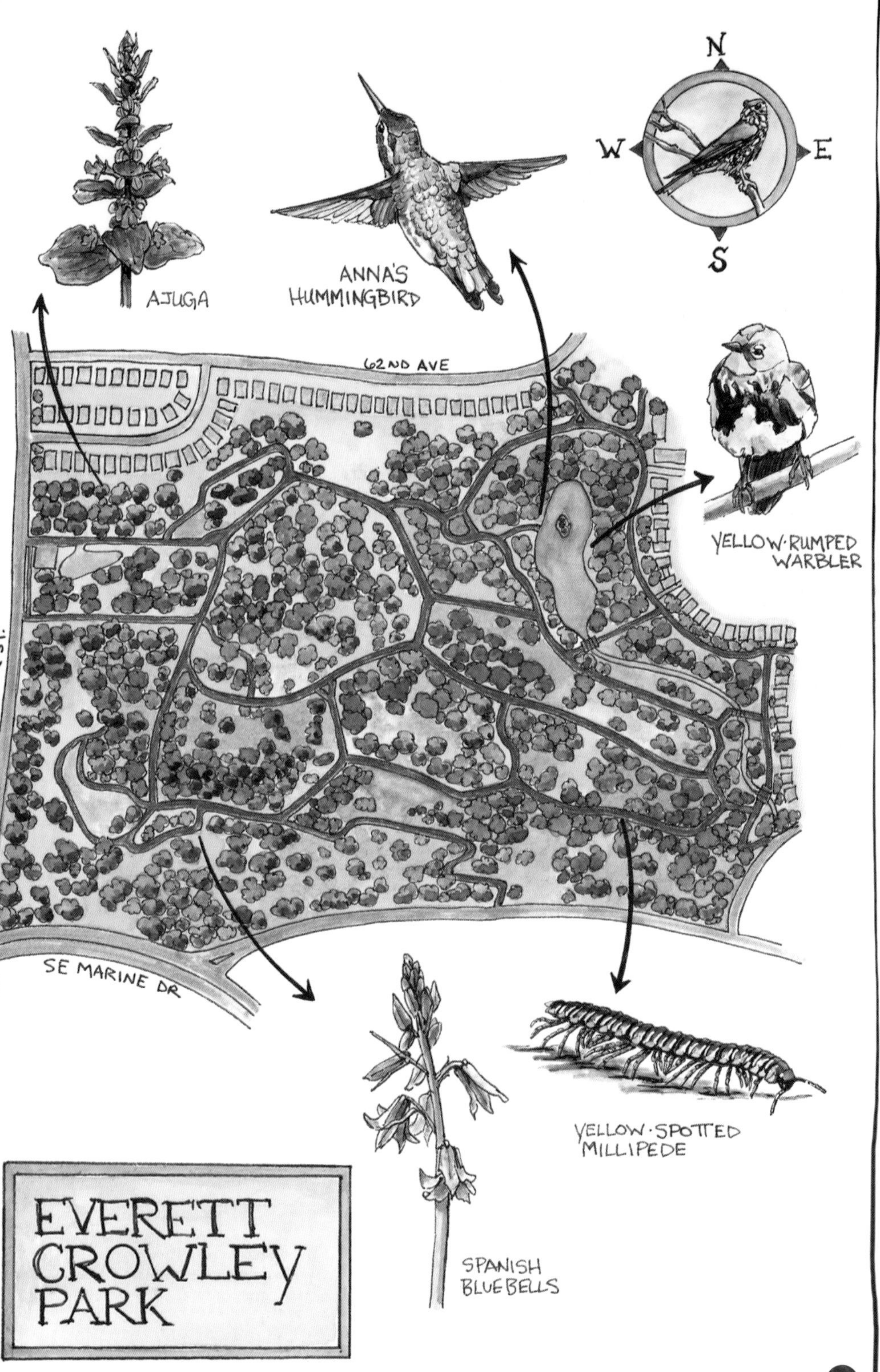

EVERETT CROWLEY PARK

Hooded Mergansers

(*Lophodytes cucullatus*)
November 21, 2021

Mostly mallards on Avalon Pond today, but a few mergansers were fishing. Hard to miss the bold patterns on the male and such a joy to draw. Hooded mergansers are the smallest mergansers in British Columbia. Feeding mainly on small fish, they dive to catch their prey and can stay underwater for up to two minutes.

Old Man's Beard

(*Clematis vitalba*)
November 21, 2021

The afternoon sun behind these seed head pods captured my attention. Intricate and mesmerizing, they shine like bright white stars. A beautiful but invasive species of woody vine, each feathery single seed (achene) can catch the wind and travel to new locations.

- Introduced from Europe and Southwest Asia.
- This woody vine is considered to be a noxious weed.
- Old Man's Beard can grow to 30.5 metres (100 feet) in length, blanketing trees and shrubs – starving them of sunlight. The weight becomes so intense, one vine can take down a mature tree.

Indian Plum

(*Oemleria cerasiformis*)
March 20, 2022

Indian plum is native to British Columbia and is an important food source for wildlife. One of the earliest plants to bloom in the spring, Indian plum is pollinated by bees, hummingbirds, and butterflies. Fruits are popular with cedar waxwings, robins, deer, and other mammals. Crushed leaves smell like cucumber, while some say the blossoms smell like watermelon rind and cat urine – likely why another common name is skunk bush.

Blue-Eyed Darner

(*Rhionaeschna multicolor*)
June 12, 2021

The earliest dragonfly fossils have been dated to 250 million years ago. Dragonfly wings have evolved to make them the most aerobatic and agile insect hunter in the world. Prey – typically mosquitos, midges, and flies – are caught in midair. Dragonflies can fly up to 55 kph (34 mph), hover, fly backwards, and even upside-down. Twenty-three out of eighty-seven species in British Columbia are rare or considered at risk of extinction.

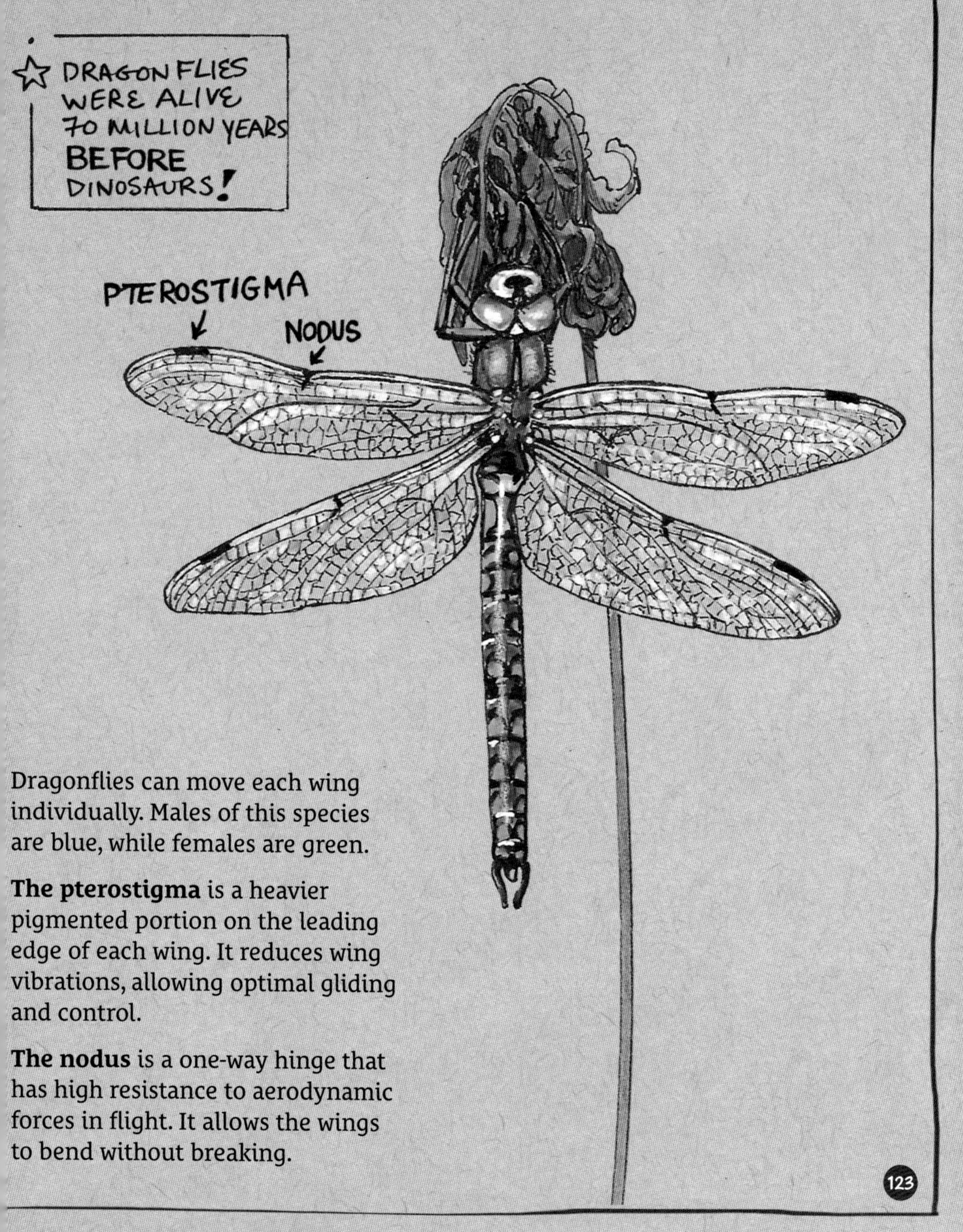

Dragonflies can move each wing individually. Males of this species are blue, while females are green.

The pterostigma is a heavier pigmented portion on the leading edge of each wing. It reduces wing vibrations, allowing optimal gliding and control.

The nodus is a one-way hinge that has high resistance to aerodynamic forces in flight. It allows the wings to bend without breaking.

Anna's Hummingbird

(*Calypte anna*)
Spring 2021

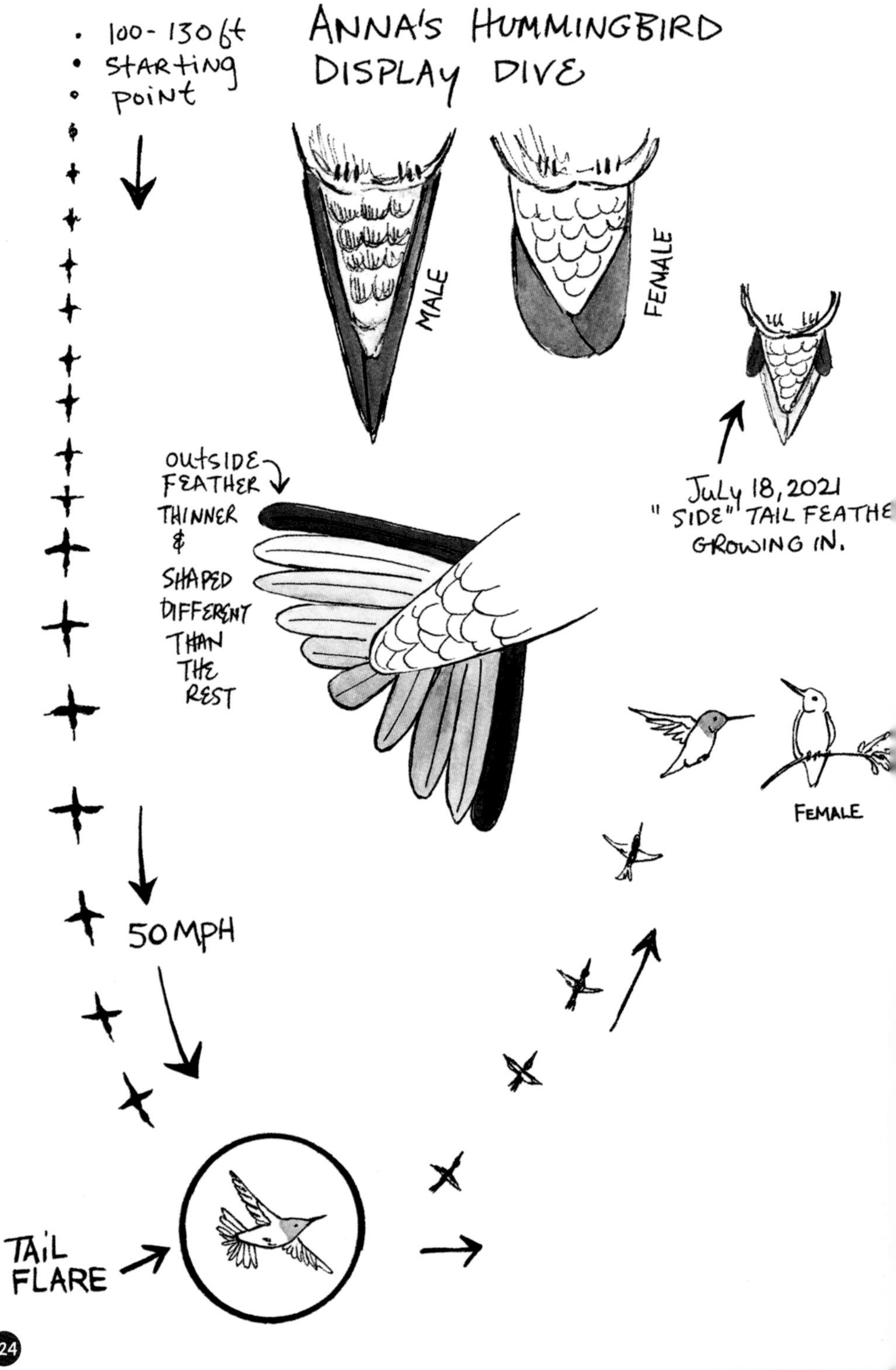

ANNA'S HUMMINGBIRD DISPLAY DIVE

100-130 ft starting point

MALE

FEMALE

OUTSIDE FEATHER THINNER & SHAPED DIFFERENT THAN THE REST

JULY 18, 2021 "SIDE" TAIL FEATHE GROWING IN.

FEMALE

50 MPH

TAIL FLARE

If you hear a high-pitched chirp while a male Anna's hummingbird is performing his aerial display dive, his tail feathers are "singing."

The outermost feather on each side of the male's tail is narrower and a different shape than the rest. At the bottom of his dive, he flares his tail for a millisecond, causing those feathers to vibrate like a clarinet reed. This makes a loud "chirp" before he pulls up in front of his target – usually a female.

Garry Oak

(*Quercus garryanna*)
Fall 2017

Garry oak leaves look and feel a bit leathery in the fall. Their intricate vein pattern reminds me of an aerial view of streets and alleyways in the city. Appropriate, in a way, because these veins provide structure for the leaf and are the "highways" that transport nutrients, water, and energy to the rest of the plant.

The more closely I studied them, the more intriguing they became. Using watercolour dry brush techiques on vellum, this studio painting allowed me to capture the complexity, textures, and subtle colour variations in each leaf.

Garry oaks, along with their associated ecosystems, are among the most rare and endangered in Canada. These habitats occupy only a small portion of the Coastal Douglas-fir zone, which itself makes up only 0.3% of the land area in British Columbia.

Garry oak ecosystems are home to more plant species than any other ecosystem in coastal British Columbia – and many of these species occur nowhere else in the country. Over one hundred species of plants, animals, birds, and insects are officially listed as "at risk" and in danger of disappearing in the wild.

Hastings Park

The Douglas Road from New Westminster to what is now New Brighton Beach Park was completed in 1865. At that time, a small settlement was established, originally known as the End of the Road and afterwards as Brighton, until it was officially named Hastings in 1868.

In 1889, the Province of British Columbia granted nearby land to the City of Vancouver "for the use, recreation, and enjoyment of the public." Hastings Park is now Vancouver's second-largest park at 62 hectares (154 acres). It includes the Pacific National Exhibition grounds, Playland, and Hastings Park Racecourse. Horse racing began here in 1892 and continues today. The park area was also used for military purposes during WWI and WWII. The Hastings Sunrise community lobbied the City and the Vancouver Board of Parks and Recreation to restore green space within the park. In 1993, the Momiji Japanese Gardens opened to the public and commemorates eight thousand Canadians of Japanese ancestry who were detained within Hastings Park at the beginning of WWII.

In 1996, an additional four hectares (9.9 acres) of the park were dedicated for nature restoration, including the Hastings Sanctuary with a pond stocked with rainbow trout and the Italian Garden created by the Italian Canadian community. The Sanctuary opened to the public in 1999 and is now considered a birding "hotspot." In 2011, a master plan for the area was released, including an ecological component to daylight an existing stream connecting Sanctuary Pond to the newly re-established saltwater marsh at New Brighton Park on Burrard Inlet. This will create an important riparian corridor and greenway.

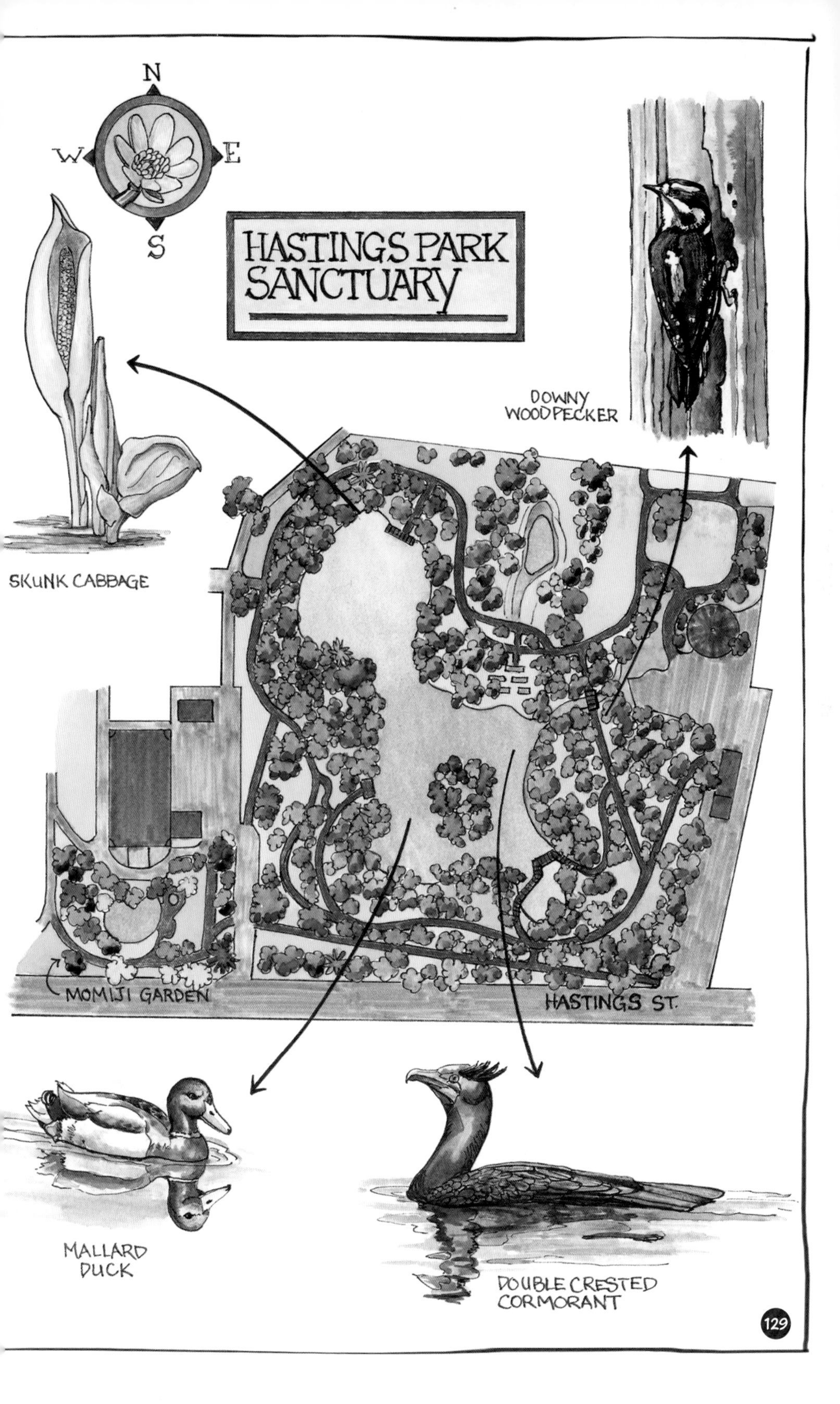
N
W
E
S
HASTINGS PARK SANCTUARY
SKUNK CABBAGE
DOWNY WOODPECKER
MOMIJI GARDEN
HASTINGS ST.
MALLARD DUCK
DOUBLE CRESTED CORMORANT

A Birding "Hotspot"

March 31, 2022

On my first visit to Hastings Park Santuary, I was thrilled to see so many different species of birds amidst the wandering pathways and wooden bridges. I then learned it is considered a birding "hotspot" – and it is definitely worthy of this title!

Even though Hastings Park is close to Burrard Inlet, I was surprised to see double-crested cormorants on the pond. Their jewel-like turquoise eyes are stunning! I was also delighted to witness this male's courtship dance. He was definitely showing off for his mate – tossing, swimming, and flying around the pond with his stick of choice for at least thirty minutes.

Male cormorants choose a nesting site, then perform courtship dances – including presenting sticks and other nesting material to the female. Nests are found on the ground or in the upper branches of trees, but are always adjacent to water. Nesting material usually consists of plant stalks, sticks, and weeds but sometimes rope, small buoys, fishnet, and plastic debris are also used.

Yellow-Rumped Warbler

(*Setophaga coronata*)

(Setophagus Coronata)
"moth eating" "crowned"

↖"AUDUBON'S" WARBLER = WEST OF THE ROCKIES

"MYRTLE" WARBLER = EASTERN N. AM. (white throats)

Lots of birds in the park today!

- Double-crested cormorants (4)
- American widgeons (2)
- Mallards (20+)
- Bushtits (6)
- Chickadees (5)
- Kinglets (6)
- Anna's hummingbird (1)
- Yellow-rumped warbler (1)
- Downy woodpecker (1)
- Cooper's hawk (1)
- Bald eagle (1)
- Red-winged blackbirds (2)
- American crows (2)
- Great blue heron (1)

Double-Crested Cormorant

(*Nannopterum auritum*)

March 31, 2022

- Double crests are only on adults & only visible during the breeding season.
- Crests look like a bad toupee!
- ALASKAN cormorants have white crests.

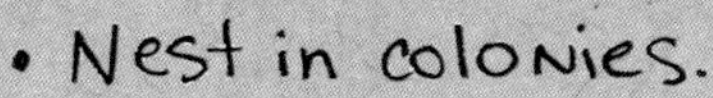

:lassic feather-drying pose

:ormorants don't have as
nuch preen oil compared to
ɔther birds, so feathers get
vaterlogged easily. This helps
hem stay underwater longer
o hunt, but they need to
pend much time drying out.

Downy Woodpecker

(*Dryobates pubescens*)
April 1, 2021

You often hear woodpeckers before you see them. Constant drumming of this female downy woodpecker helped me find her as she searched for a snack.

Red Dead-Nettle

(*Lamium purpureum*)
March 21, 2022

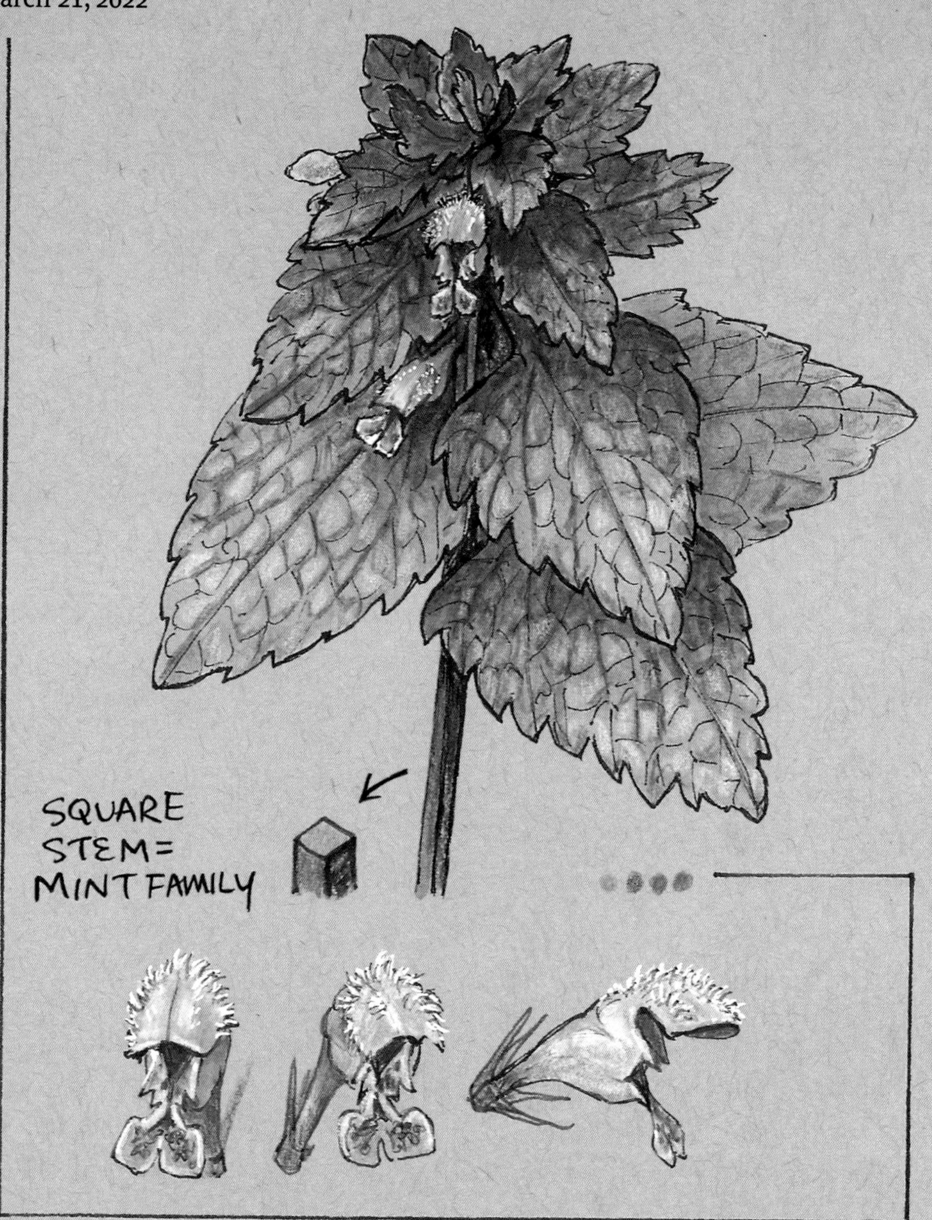

This small plant is easy to walk by without a second look, but a closer inspection reveals tiny lavender blossoms with tons of character.

- Also known as purple archangel and purple dead-nettle.
- Flowers are rich in nectar and pollen. Bees love them!
- Dried leaves have been used as a poultice to stop bleeding. Fresh leaves applied topically assist the body in healing cuts and bruises.

Bee Fly

(*Bombyliidae family*)
August 20, 2021

Also known as the humble fly, on close inspection the bee fly lives up to its name, looking like an equal cross between a bee and a fly. There are at least 4,500 species, so it was surprising to learn that they are not well studied. I was also surprised to learn that their tongue (proboscis) cannot retract! It stays extended long and straight, perfect for drinking nectar from flowers with long floral tubes.

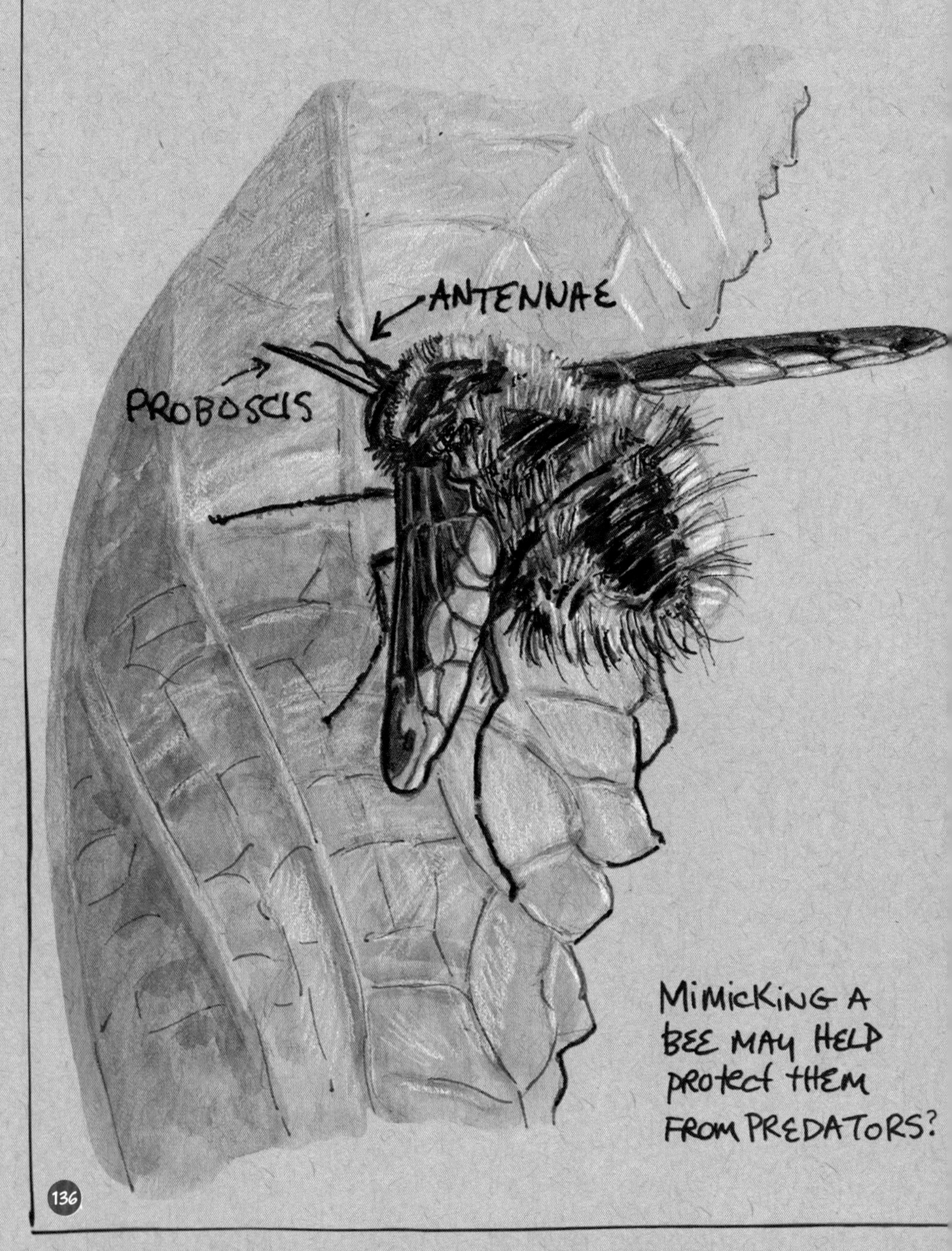

Townsend's Warbler

(*Setophaga townsendi*)
October 27, 2021

At first you only see flashes of yellow dancing through the trees as warblers zip in and out, catching insects. A few species migrate through Vancouver in the spring and fall. The Townsend's warbler winters in Central America and Mexico and travels as far as Alaska to breed. It is always a delight to see them.

Creating Your Own Nature Sketchbook

Starting Your Nature Sketchbook Journal

Starting a nature sketchbook journal is easy. No special materials are needed, just a pencil or pen and some paper will work just fine. If you happen to have a notebook, journal, diary, or day planner, that's even better. One important tip is to choose a book size that is easy to carry. Anything too large or cumbersome usually gets left at home. If you want to invest in coloured pencils, markers, or small watercolour paint kit, adding colour can be rewarding.

There is no right or wrong way to work in a sketchbook journal. This is your own nature adventure. When you go outside – look, explore, and write questions in your journal. Watch and record what you see, what you hear, and what catches your attention. Slow down, watch, and listen to what is around you. Think of nature like a big treasure map waiting to be discovered. Then use your journal to document your journey and the stories you uncover.

Nature journaling centres around being curious and asking questions, then putting your discoveries on the page. Journaling is not about making "pretty pictures" – it's a way to connect meaningfully with nature wherever you are and create lasting memories of your experiences.

Here are a few helpful steps to get started:

1. Schedule "green time" dates in your calendar. You can spend as much or as little time as you have available. A lunch or coffee break, a walk home from work or school, or a dedicated time in your favourite park are all perfect opportunities to put observations, notes, and sketches on your pages.

2. Collect your journal materials and then go outside! If you cannot get outside or the weather is uncooperative, look out a window and see what you notice. Perhaps the shapes or types of clouds, the wildlife on the ground or in trees, or your impressions of rain, wind, or snow.

3. Don't be intimidated by a blank page. To get started, write down the date, time, and location of where you are. This is called metadata and including it on your page turns what you experience that day into a historical, scientific record. It puts the stories you encounter into context within the bigger nature picture. You can even add information like the weather, temperature, wind direction, and phases of the moon.

4. Be curious. What do you notice? Hear? Smell? An interesting texture; birds singing or flitting nearby; a lovely plant, seedpod, or insect; the scent of pine? Use all your senses. If you are more comfortable writing, jot down notes or create a poem. If you are more comfortable drawing, sketch what you see. Words, labels, and pictures work great in combination. While sketchbook pages can be "beautiful" and "artistic," adding words shifts your focus from making a "perfect" drawing to becoming involved in the story in front of you. Sketches can have as much or as little detail as you want.

5. Dress for the weather and changing conditions and take a snack if you will be out for a few hours. When you are comfortable, it's easier to slow down and be still.

6. If you have a small magnifying glass and/or binoculars, then pack them along. Both tools help you get a better look at things not easily seen with the naked eye.

7. Don't worry if you can't identify a plant, animal, or insect. If you have a smart phone, apps like Seek and iNaturalist can help you ID species in the field. If not, that's okay too! Get as much information as you can down on the page – the colour, size, habitat, any details you see – then research what you've found later when you return home.

8. Start small. If you are not sure what to draw, select a leaf to sketch rather than the whole tree. If you have coloured pencils or a paint kit, capture your surroundings with a row of coloured swatches matching what you see.

9. Tread lightly! Turning over rocks and logs can reveal incredible discoveries that are wonderful to add to your journal. Please replace them when you are done. These are often critical shelter and habitats for organisms that live there – many that you may not even notice.

10. Draw things where you find them. It may be tempting, but don't pick flowers or uproot plants. In many areas, this is even illegal. If you have limited time or the weather is uninviting and you have a camera or smartphone, you can take reference photos to fill in details when you get home.

11. If you get stuck on what to add on your page, remember to ask questions using the guiding principles: "I notice, I wonder, and it reminds me of."

12. Most of all, have fun!

"Look deep into nature and you will understand everything better."

— ALBERT EINSTEIN, physicist

Naturehoods: Finding Nature in Your Neighbourhood

Nature is everywhere, but using a few, helpful techniques can increase your chance of seeing birds, animals, and insects.

- Always respect your surroundings and any wildlife you encounter – stay quiet and move slowly. Wildlife tends to go about their business when they don't realize you're there!
- If you can find a safe place to sit down, get comfortable and take a deep breath. Close your eyes and engage all your senses. Take note of different sounds and what direction they are coming from. Then open your eyes and make notes about what you notice. By expanding your awareness, you become more attuned to subtle activities of nature around you.
- When you plan outings, include areas that contain ecotones. These are transitional areas located between two different types of ecosystems. For example, riverbanks between a forest and river, high salt marshes between an ocean and marshland, hedgerows between meadows and woodland, or estuaries between fresh and salt water tend to support higher numbers of species and are magical areas to view wildlife.
- Keep an appropriate distance from wildlife and leave before you cause them any distress. Take notice if behaviour shifts – if they stop what they are doing or begin to act nervous – be respectful and move on. Depending on the time of year, birds (especially) may be reluctant to leave an area if they have a nest or young nearby. Keep a good distance between you and any larger mammals you may encounter, including raccoons, deer, and coyotes.
- If you find a nest, keep your distance. Some birds will abandon their young if they feel their nest has been "discovered" or they feel threatened. Any human scent that is left behind can also lead predators to the eggs and babies.
- Be careful if you trim trees and hedges on your property in spring and summer. Thick leaf cover is ideal to camouflage and protect nests and baby birds.
- Journal responsibly – don't leave anything behind except footprints! Be sure to pack out any food or snack wrappers and any food scraps left if you have a meal. If you happen to see discarded garbage, be an environmental hero and pack it out if you are able and can do so safely. The wild residents will thank you.
- Stay on designated trails and look high and low. As you spend more time in nature, your brain, eyes and ears become more aware of what's "normal" – the way the wind blows in the trees, the patterns of leaves and branches on shrubs, how plants grow, and common birdsongs. Then the things that might be out of place are more apparent. For example, a dark shape on a high branch, an unusual sound, a unique birdcall, or leaves rustling in an odd way on a still day will catch your attention. These are all signs there is something interesting to find!

- By visiting the same location regularly and at different times of the year, you start to notice how that space changes with the seasons.
- Don't feed the wildlife! Eating "human food" is often detrimental and sometimes fatal for birds and animals.
- Remember to bring your journal and supplies! When you are quiet and involved in sketching and writing notes, wildlife gets more comfortable with your presence and often carries on with their daily routines making it easier to capture their stories on paper.

"I took a walk in the woods and came out taller than the trees."

— HENRY DAVID THOREAU
naturalist, poet, philosopher

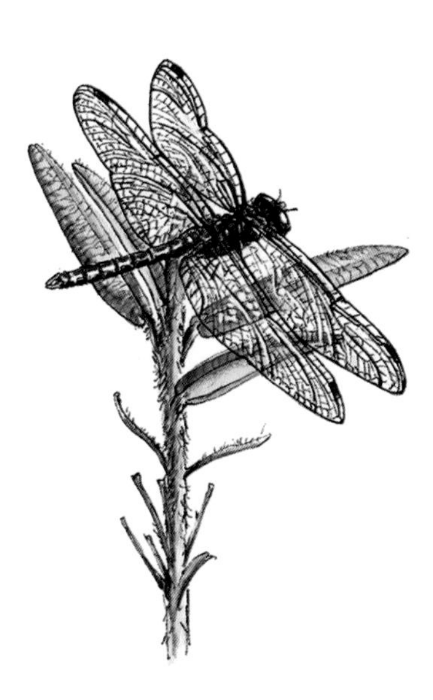

Further Reading

The Laws Guide to Nature Drawing and Journaling by John Muir Laws. Berkeley, CA: Heyday, 2016. ISBN 9781597143158.

Creating Textures in Pen & Ink with Watercolor by Claudia Nice. Cincinnati: North Light Books, 2012. ISBN 9781581807257.

The Laws Guide to Drawing Birds by John Muir Laws. Berkeley, CA: Heyday, 2012. ISBN 9781597141956.

Natural History Painting with the Eden Project by Rosie Martin and Meriel Thurstan. London: Batsford, 2009. ISBN 9781906388492.

Artist's Journal Workshop: Creating Your Life in Words and Pictures by Cathy Johnson. Cincinnati: North Light Books, 2011. ISBN 9781440308680.

Keeping a Nature Journal by Clare Walker Leslie and Charles E. Roth. North Adams, MA: Storey Publishing, 2003. ISBN 9781580174930.

Painting Nature in Watercolor by Cathy Johnson. Cincinnati: North Light Books, 2014. ISBN 9781440328831.

Acknowledgments

I have many heroes. First, this book is dedicated to my Mom with love. Thank you for introducing me to the magic of nature! For letting me run barefoot in the woods with our dog, build tree forts and snow elephants, and for instilling an everlasting respect for wild things. Most of all, for instilling the belief in me that anything is possible.

I thank my dear friends for their unwavering support and encouragement during the process of writing and preparing the images for this book. Christina Bolzner, Anne-Marie Comte, Mary Landell, Anne Janeda, Angela Muellers, Judith Joseph, and Karim Khan – I could not have done this without you.

I also want to acknowledge and thank the global community of nature artists and sketchbook journalers who are helping spread the word about the benefits of keeping a regular sketchbook practice and who generously share their techniques and talents with the world, most notably John Muir Laws.

And a special thank you to Midtown Press and the team of Louis Anctil, Ann-Marie Metten, and Denis Hunter, who accepted this project and guided me on the journey to this finished product.

Index of Plants, Birds, and Animals

About the Author

Natural science, medical, and botanical illustrator Vicky Earle delights in connecting people to nature through art. After graduating from the University of Toronto with a degree in medical illustration and biocommunications, Vicky moved to the West Coast and has called Vancouver home for over three decades. She worked in the health care profession for thirty years, and is now dedicated to sharing natural history stories that take place every day all around us. Vicky's goal is to help foster curiosity and an affinity for the natural world by encouraging people to follow their own art practice. Her preferred medium is watercolour on paper, occasionally using metalpoint and mixed media. She has kept sketchbooks throughout her career and began a dedicated sketchbook journal practice in 2014.

Vicky's work has been part of numerous international juried exhibitions, including *Focus on Nature* at the New York State Museum. She represented Canada during the *Botanical Art Worldwide* exhibition, a groundbreaking collaboration between botanical artists, organizations, and global institutions that created and exhibited botanical artworks of native plants found in each of twenty-five participating countries. Vicky is a member of Nature Vancouver, the Guild of Natural Science Illustrators, the American Society of Botanical Artists, and is a Signature member of the Artists for Conservation.

For updates on current projects and artwork, visit her website *drawinnature.com* and follow her on Instagram *@drawinnature*.